This report contains the collective views of a[n] experts and does not necessarily represent th[e] of the United Nations Environment Programm[e], Organization or the World Health Organization.

Environmental Health Criteria 226

PALLADIUM

First draft prepared by Dr Christine Melber, Dr Detlef Keller and Dr Inge Mangelsdorf, Fraunhofer Institute for Toxicology and Aerosol Research, Hanover, Germany

Published under the joint sponsorship of the United Nations Environment Programme, the International Labour Organization and the World Health Organization, and produced within the framework of the Inter-Organization Programme for the Sound Management of Chemicals.

World Health Organization
Geneva, 2002

The **International Programme on Chemical Safety (IPCS)**, established in 1980, is a joint venture of the United Nations Environment Programme (UNEP), the International Labour Organization (ILO) and the World Health Organization (WHO). The overall objectives of the IPCS are to establish the scientific basis for assessment of the risk to human health and the environment from exposure to chemicals, through international peer review processes, as a prerequisite for the promotion of chemical safety, and to provide technical assistance in strengthening national capacities for the sound management of chemicals.

The **Inter-Organization Programme for the Sound Management of Chemicals (IOMC)** was established in 1995 by UNEP, ILO, the Food and Agriculture Organization of the United Nations, WHO, the United Nations Industrial Development Organization, the United Nations Institute for Training and Research and the Organisation for Economic Co-operation and Development (Participating Organizations), following recommendations made by the 1992 UN Conference on Environment and Development to strengthen cooperation and increase coordination in the field of chemical safety. The purpose of the IOMC is to promote coordination of the policies and activities pursued by the Participating Organizations, jointly or separately, to achieve the sound management of chemicals in relation to human health and the environment.

WHO Library Cataloguing-in-Publication Data

Palladium.

(Environmental health criteria ; 226)

1.Palladium - toxicity 2.Palladium - adverse effects 3.Environmental exposure 4.Occupational exposure 5.Risk assessment I.International Programme for Chemical Safety II.Series

ISBN 92 4 157226 4   (NLM classification: QV 290)
ISSN 0250-863X

The World Health Organization welcomes requests for permission to reproduce or translate its publications, in part or in full. Applications and enquiries should be addressed to the Office of Publications, World Health Organization, Geneva, Switzerland, which will be glad to provide the latest information on any changes made to the text, plans for new editions, and reprints and translations already available.

The designations employed and the presentation of the material in this publication do not imply the expression of any opinion whatsoever on the part of the Secretariat of the World Health Organization concerning the legal status of any country, territory, city or area or of its authorities, or concerning the delimitation of its frontiers or boundaries.

The mention of specific companies or of certain manufacturers' products does not imply that they are endorsed or recommended by the World Health Organization in preference to others of a similar nature that are not mentioned. Errors and omissions excepted, the names of proprietary products are distinguished by initial capital letters.

The Federal Ministry for the Environment, Nature Conservation and Nuclear Safety, Germany, provided financial support for, and undertook the printing of, this publication.

CONTENTS

ENVIRONMENTAL HEALTH CRITERIA FOR PALLADIUM

NOTE TO READERS OF THE CRITERIA MONOGRAPHS

Every effort has been made to present information in the criteria monographs as accurately as possible without unduly delaying their publication. In the interest of all users of the Environmental Health Criteria monographs, readers are requested to communicate any errors that may have occurred to the Director of the International Programme on Chemical Safety, World Health Organization, Geneva, Switzerland, in order that they may be included in corrigenda.

* * *

A detailed data profile and a legal file can be obtained from the International Register of Potentially Toxic Chemicals, Case postale 356, 1219 Châtelaine, Geneva, Switzerland (telephone no. + 41 22 – 9799111, fax no. + 41 22 – 7973460, E-mail irptc@unep.ch).

* * *

This publication was made possible by grant number 5 U01 ES02617-15 from the National Institute of Environmental Health Sciences, National Institutes of Health, USA, and by financial support from the Federal Ministry for the Environment, Nature Conservation and Nuclear Safety, Germany.

Environmental Health Criteria

PREAMBLE

Objectives

In 1973, the WHO Environmental Health Criteria Programme was initiated with the following objectives:

(i) to assess information on the relationship between exposure to environmental pollutants and human health, and to provide guide-lines for setting exposure limits;

(ii) to identify new or potential pollutants;

(iii) to identify gaps in knowledge concerning the health effects of pollutants;

(iv) to promote the harmonization of toxicological and epidemio-logical methods in order to have internationally comparable results.

The first Environmental Health Criteria (EHC) monograph, on mercury, was published in 1976, and since that time an ever-increasing number of assessments of chemicals and of physical effects have been produced. In addition, many EHC monographs have been devoted to evaluating toxicological methodology, e.g., for genetic, neurotoxic, teratogenic and nephrotoxic effects. Other publications have been concerned with epidemiological guidelines, evaluation of short-term tests for carcinogens, biomarkers, effects on the elderly and so forth.

Since its inauguration, the EHC Programme has widened its scope, and the importance of environmental effects, in addition to health effects, has been increasingly emphasized in the total evaluation of chemicals.

The original impetus for the Programme came from World Health Assembly resolutions and the recommendations of the 1972 UN Conference on the Human Environment. Subsequently, the work became an integral part of the International Programme on Chemical

Safety (IPCS), a cooperative programme of UNEP, ILO and WHO. In this manner, with the strong support of the new partners, the importance of occupational health and environmental effects was fully recognized. The EHC monographs have become widely established, used and recognized throughout the world.

The recommendations of the 1992 UN Conference on Environment and Development and the subsequent establishment of the Intergovernmental Forum on Chemical Safety with the priorities for action in the six programme areas of Chapter 19, Agenda 21, all lend further weight to the need for EHC assessments of the risks of chemicals.

Scope

The criteria monographs are intended to provide critical reviews on the effects on human health and the environment of chemicals and of combinations of chemicals and physical and biological agents. As such, they include and review studies that are of direct relevance for the evaluation. However, they do not describe *every* study carried out. Worldwide data are used and are quoted from original studies, not from abstracts or reviews. Both published and unpublished reports are considered, and it is incumbent on the authors to assess all the articles cited in the references. Preference is always given to published data. Unpublished data are used only when relevant published data are absent or when they are pivotal to the risk assessment. A detailed policy statement is available that describes the procedures used for unpublished proprietary data so that this information can be used in the evaluation without compromising its confidential nature (WHO (1999) Revised Guidelines for the Preparation of Environmental Health Criteria Monographs. PCS/99.9, Geneva, World Health Organization).

In the evaluation of human health risks, sound human data, whenever available, are preferred to animal data. Animal and *in vitro* studies provide support and are used mainly to supply evidence missing from human studies. It is mandatory that research on human subjects is conducted in full accord with ethical principles, including the provisions of the Helsinki Declaration.

The EHC monographs are intended to assist national and international authorities in making risk assessments and subsequent risk management decisions. They represent a thorough evaluation of risks and are not, in any sense, recommendations for regulation or standard setting. These latter are the exclusive purview of national and regional governments.

Content

The layout of EHC monographs for chemicals is outlined below.

* Summary — a review of the salient facts and the risk evaluation of the chemical
* Identity — physical and chemical properties, analytical methods
* Sources of exposure
* Environmental transport, distribution and transformation
* Environmental levels and human exposure
* Kinetics and metabolism in laboratory animals and humans
* Effects on laboratory mammals and *in vitro* test systems
* Effects on humans
* Effects on other organisms in the laboratory and field
* Evaluation of human health risks and effects on the environment
* Conclusions and recommendations for protection of human health and the environment
* Further research
* Previous evaluations by international bodies, e.g., IARC, JECFA, JMPR

Selection of chemicals

Since the inception of the EHC Programme, the IPCS has organized meetings of scientists to establish lists of priority chemicals for subsequent evaluation. Such meetings have been held in Ispra, Italy, 1980; Oxford, United Kingdom, 1984; Berlin, Germany, 1987; and North Carolina, USA, 1995. The selection of chemicals has been based on the following criteria: the existence of scientific evidence that the substance presents a hazard to human health and/or the environment; the possible use, persistence, accumulation or degradation of the substance shows that there may be significant human or environmental exposure; the size and nature of populations at risk (both human and other species) and risks for the environment; international concern, i.e., the substance is of major interest to several countries; adequate data on the hazards are available.

If an EHC monograph is proposed for a chemical not on the priority list, the IPCS Secretariat consults with the cooperating organizations and all the Participating Institutions before embarking on the preparation of the monograph.

Procedures

The order of procedures that result in the publication of an EHC monograph is shown in the flow chart on the next page. A designated staff member of IPCS, responsible for the scientific quality of the document, serves as Responsible Officer (RO). The IPCS Editor is responsible for layout and language. The first draft, prepared by consultants or, more usually, staff from an IPCS Participating Institution, is based initially on data provided from the International Register of Potentially Toxic Chemicals and from reference databases such as Medline and Toxline.

The draft document, when received by the RO, may require an initial review by a small panel of experts to determine its scientific quality and objectivity. Once the RO finds the document acceptable as a first draft, it is distributed, in its unedited form, to well over 150 EHC contact points throughout the world who are asked to comment on its completeness and accuracy and, where necessary, provide additional material. The contact points, usually designated by governments, may be Participating Institutions, IPCS Focal Points or individual scientists known for their particular expertise. Generally, some four months are allowed before the comments are considered by the RO and author(s). A second draft incorporating comments received and approved by the Director, IPCS, is then distributed to Task Group members, who carry out the peer review, at least six weeks before their meeting.

The Task Group members serve as individual scientists, not as representatives of any organization, government or industry. Their function is to evaluate the accuracy, significance and relevance of the information in the document and to assess the health and environmental risks from exposure to the chemical. A summary and recommendations

EHC PREPARATION FLOW CHART

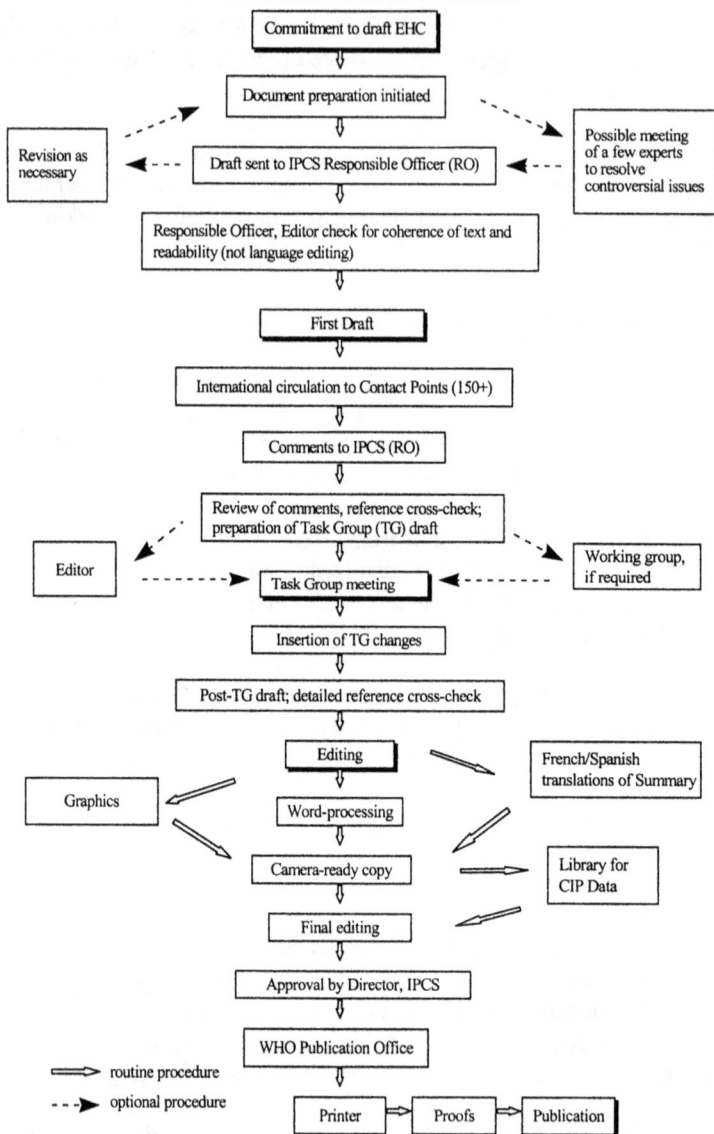

Commitment to draft EHC

Document preparation initiated

Revision as necessary ← Draft sent to IPCS Responsible Officer (RO) → Possible meeting of a few experts to resolve controversial issues

Responsible Officer, Editor check for coherence of text and readability (not language editing)

First Draft

International circulation to Contact Points (150+)

Comments to IPCS (RO)

Review of comments, reference cross-check; preparation of Task Group (TG) draft

Editor ← Task Group meeting → Working group, if required

Insertion of TG changes

Post-TG draft; detailed reference cross-check

Editing → French/Spanish translations of Summary

Graphics ← Word-processing

Camera-ready copy → Library for CIP Data

Final editing

Approval by Director, IPCS

WHO Publication Office

⇒ routine procedure
--→ optional procedure

Printer ⇒ Proofs ⇒ Publication

for further research and improved safety aspects are also required. The composition of the Task Group is dictated by the range of expertise required for the subject of the meeting and by the need for a balanced geographical distribution.

The three cooperating organizations of the IPCS recognize the important role played by nongovernmental organizations. Representatives from relevant national and international associations may be invited to join the Task Group as observers. While observers may provide a valuable contribution to the process, they can speak only at the invitation of the Chairperson. Observers do not participate in the final evaluation of the chemical; this is the sole responsibility of the Task Group members. When the Task Group considers it to be appropriate, it may meet *in camera*.

All individuals who as authors, consultants or advisers participate in the preparation of the EHC monograph must, in addition to serving in their personal capacity as scientists, inform the RO if at any time a conflict of interest, whether actual or potential, could be perceived in their work. They are required to sign a conflict of interest statement. Such a procedure ensures the transparency and probity of the process.

When the Task Group has completed its review and the RO is satisfied as to the scientific correctness and completeness of the document, the document then goes for language editing, reference checking and preparation of camera-ready copy. After approval by the Director, IPCS, the monograph is submitted to the WHO Office of Publications for printing. At this time, a copy of the final draft is sent to the Chairperson and Rapporteur of the Task Group to check for any errors.

It is accepted that the following criteria should initiate the updating of an EHC monograph: new data are available that would substantially change the evaluation; there is public concern for health or environmental effects of the agent because of greater exposure; an appreciable time period has elapsed since the last evaluation.

All Participating Institutions are informed, through the EHC progress report, of the authors and institutions proposed for the drafting of the documents. A comprehensive file of all comments received on drafts of each EHC monograph is maintained and is available on request. The Chairpersons of Task Groups are briefed before each meeting on their role and responsibility in ensuring that these rules are followed.

WHO TASK GROUP ON ENVIRONMENTAL HEALTH CRITERIA FOR PALLADIUM

Members

Professor Werner Aberer, Department of Dermatology, University of Graz, Graz, Austria

Dr Janet Kielhorn, Fraunhofer Institute for Toxicology and Aerosol Research, Hanover, Germany (*Co-Rapporteur*)

Assistant Professor Patrick Koch, Department of Dermatology, Saarland University Hospital, Homburg/Saar, Germany

Dr Jorma Maki-Paakkanen, Department of Environmental Medicine, National Public Health Institute, Kuopio, Finland

Mr Heath Malcolm, Centre for Ecology and Hydrology, Monks Wood, Abbots Ripton, Huntingdon, United Kingdom (*Co-Rapporteur*)

Professor Gunnar Nordberg, Unit of Environmental Medicine, Department of Public Health and Clinical Medicine, Umea University, Umea, Sweden (*Chairman*)

Professor John C. Wataha, Department of Oral Rehabilitation, School of Dentistry, Medical College of Georgia, Augusta, Georgia, USA

Dr Mark White, Health and Safety Laboratory, Sheffield, United Kingdom

Secretariat

Mr Yoshikazu Hayashi, International Programme on Chemical Safety, World Health Organization, Geneva, Switzerland

Dr Detlef Keller, Fraunhofer Institute for Toxicology and Aerosol Research, Hanover, Germany

Dr Inge Mangelsdorf, Fraunhofer Institute for Toxicology and Aerosol Research, Hanover, Germany

Dr Christine Melber, Fraunhofer Institute for Toxicology and Aerosol Research, Hanover, Germany

Professor Fedor Valić, WHO/IPCS Scientific Adviser, Department of Environmental and Occupational Health, Andrija Štampar School of Public Health, University of Zagreb, Zagreb, Croatia (*Responsible Officer*)

Observer

Dr Peter Linnett, Johnson Matthey plc, Royston, United Kingdom

ENVIRONMENTAL HEALTH CRITERIA FOR PALLADIUM

A Task Group on Environmental Health Criteria for Palladium met at the Fraunhofer Institute for Toxicology and Aerosol Research, Hanover, Germany, from 8 to 12 May 2000. Professor H. Muhle, Deputy Director, Fraunhofer Institute for Toxicology and Aerosol Research, opened the meeting and welcomed the participants on behalf of the host institution. Professor F. Valić welcomed the participants on behalf of the Director, IPCS, and the heads of the three cooperating organizations of the IPCS (UNEP/ILO/WHO). The Task Group reviewed and revised the draft of the monograph, made an evaluation of the risks for human health and the environment from exposure to palladium and made recommendations for health protection and further research.

The first draft was prepared by Dr Christine Melber, Dr Detlef Keller and Dr Inge Mangelsdorf, Fraunhofer Institute for Toxicology and Aerosol Research, Hanover, Germany. The second draft was also prepared by the same authors, incorporating comments received following the circulation of the first draft to the IPCS Contact Points for Environmental Health Criteria monographs.

Professor F. Valić was responsible for the overall scientific content of the monograph, and Dr P.G. Jenkins, IPCS Central Unit, was responsible for coordinating the technical editing of the monograph.

The efforts of all who helped in the preparation and finalization of the monograph are gratefully acknowledged.

ACRONYMS AND ABBREVIATIONS

AAS	atomic absorption spectrometry
ATPase	adenosine triphosphatase
Bq	becquerel
Ci	curie (1 Ci = 3.7×10^{10} Bq)
DNA	deoxyribonucleic acid
EC_{50}	median effective concentration
EHC	Environmental Health Criteria monograph
FAO	Food and Agriculture Organization of the United Nations
FEV_1	forced expiratory volume in 1 s
GF-AAS	graphite furnace atomic absorption spectrometry
IARC	International Agency for Research on Cancer
IC_{50}	median inhibitory concentration
ICP	inductively coupled plasma
ICP-AES	inductively coupled plasma atomic emission spectrometry
ICP-MS	inductively coupled plasma mass spectrometry
ILO	International Labour Organization
IPCS	International Programme on Chemical Safety
JECFA	Joint FAO/WHO Expert Meeting on Food Additives
JMPR	Joint FAO/WHO Meeting on Pesticide Residues
K_i	inhibition constant
LC_{50}	median lethal concentration
LD_{50}	median lethal dose
MELISA	memory lymphocyte immunostimulation assay
MMAD	mass median aerodynamic diameter
MS	mass spectrometry
MTT	3-(4,5-dimethylthiazol-2-yl)-2,5-diphenyl tetrazolium bromide
NBS	National Bureau of Standards (USA)

NOAEL	no-observed-adverse-effect level
NOEC	no-observed-effect concentration
NOEL	no-observed-effect level
OECD	Organisation for Economic Co-operation and Development
PGM	platinum group metal
$PM_{2.5}$	particulate matter with aerodynamic diameter <2.5 µm
PM_{10}	particulate matter with aerodynamic diameter <10 µm
RNA	ribonucleic acid
RO	Responsible Officer
SRM	Standard Reference Material
TC_{50}	concentration causing 50% toxicity
UN	United Nations
UNEP	United Nations Environment Programme
US	United States
UV	ultraviolet
WHO	World Health Organization

1. SUMMARY

1.1 Identity, physical and chemical properties and analytical methods

Palladium is a steel-white, ductile metallic element resembling and occurring with the other platinum group metals (PGMs) and nickel. It exists in three states: Pd^0 (metallic), Pd^{2+} and Pd^{4+}. It can form organo-metallic compounds, only few of which have found industrial uses. Palladium metal is stable in air and resistant to attack by most reagents except aqua regia and nitric acid.

Currently, there are no published measurement methods that distinguish between different species of soluble or insoluble palladium in the environment.

Commonly used analytical methods for the quantification of palladium compounds are graphite furnace atomic absorption spectrometry and inductively coupled plasma mass spectrometry, the latter having the possibility of simultaneous multi-element analysis.

1.2 Sources of human and environmental exposure

Palladium occurs together with the other PGMs at very low concentrations (<1 µg/kg) in the Earth's crust. For industrial use, it is recovered mostly as a by-product of nickel, platinum and other base metal refining. Its separation from the PGMs depends upon the type of ore in which it is found.

Economically important sources exist in Russia, South Africa and North America. The worldwide mining of palladium is estimated to yield about 260 tonnes/year.

Palladium and its alloys are used as catalysts in the (petro)chemical and, above all, the automotive industries. Demand for palladium in automobile catalysts rose from 24 tonnes in 1993 to 139 tonnes in 1998, as palladium-rich technology was adopted in many gasoline-fuelled cars.

Applications for electronics and electrical technology include use in metallization processes (thick film paste), electrical contacts and switching systems.

Palladium alloys are also widely used in dentistry (e.g., for crowns and bridges).

Quantitative data on emissions of palladium into the atmosphere, hydrosphere or geosphere from natural or industrial sources are not available.

Automobile catalysts are mobile sources of palladium. Around 60% of European gasoline-fuelled cars sold in 1997 and also many Japanese and US cars were equipped with palladium-containing catalysts. Data on the exact palladium emission rate of cars equipped with modern monolithic palladium/rhodium three-way catalysts are still scarce. The particulate palladium released from a new palladium-containing catalyst ranged from 4 to 108 ng/km. These values are of the same order of magnitude as previously reported platinum emissions from catalysts.

1.3 Environmental transport, distribution and transformation

Most of the palladium in the biosphere is in the form of the metal or the metal oxides, which are almost insoluble in water, are resistant to most reactions in the biosphere (e.g., abiotic degradation, ultraviolet radiation, oxidation by hydroxyl radicals) and do not volatilize into air. By analogy to other PGMs, metallic palladium is not expected to be biologically transformable.

Under appropriate pH and redox potential conditions, it is assumed that peptides or humic or fulvic acids bind palladium in the aquatic environment. Palladium has been found in the ash of a number of plants, leading to the suggestion that palladium is more environmentally mobile and thus bioavailable to plants than is platinum.

1.4 Environmental levels and human exposure

In contrast to the large body of information concerning concentrations of metals such as lead or nickel in the environment, there is little information on palladium. Concentrations of palladium in surface water, where it is detected, generally range from 0.4 to 22 ng/litre (fresh water) and from 19 to 70 pg/litre (salt water). Concentrations reported in soil range from <0.7 to 47 µg/kg. These soil samples were all collected from areas near major roads.

Concentrations reported in sewage sludge range from 18 to 260 µg/kg, although a concentration of 4700 µg/kg has been reported in a sludge contaminated by discharges from the local jewellery industry. Drinking-water samples usually contain no palladium or <24 ng palladium/litre. The few data available show that palladium can be present in tissues of small aquatic invertebrates, different types of meat, fish, bread and plants.

The general population is primarily exposed to palladium through dental alloys, jewellery, food and emissions from automobile catalytic converters.

The human average dietary intake of palladium appears to be up to 2 µg/day.

In analogy to platinum, ambient air levels of palladium below 110 pg/m^3 can be expected in urban areas where palladium catalysts are used. Therefore, the inhalative palladium uptake is very low. In roadside dust, soil and grass samples, a slight accumulation of palladium has been detected, correlating with traffic density and distance from the road.

Oral exposure in the general environment is very important and may occur by daily direct contact of the gingiva with palladium dental alloys. Skin exposure may occur by contact with jewellery containing palladium.

Dental alloys are the most frequent cause of constant palladium exposure. The corrosive behaviour of palladium-containing dental alloys in the mouth can be influenced by the addition of other metals (such as copper, gallium and indium) and processing of the alloy. Palladium–copper alloys with high copper content may be less corrosion-resistant than palladium alloys with low copper content. Palladium release from palladium-containing dental restorations shows substantial individual variation depending on the dental condition, the material involved and personal habits (e.g., gum chewing). Clinical data for iatrogenic exposure are of limited value, as the few case-studies have methodological deficiencies, such as limited numbers of tissue samples and poorly matched control groups. It is, therefore, difficult to make an accurate quantitative statement regarding daily intake, and the proposed value of $\leq 1.5–15$ µg palladium/day per person thus remains a crude estimation.

There is some information on palladium levels in the general population, where levels in urine were in the range of 0.006–<0.3 µg/litre in adults.

Most occupational exposures to palladium (salts) occur during palladium refining and catalyst manufacture. There are few exposure measurements, ranging from 0.4 to 11.6 µg/m³ as an 8-h time-weighted average. No recent data are available for biological monitoring of workers exposed to palladium and its salts.

Dental technicians may be exposed to peaks of palladium dust during processing and polishing of dental casting alloys containing palladium, especially if adequate protective measures (dust extraction or aspiration techniques) are not taken.

1.5 Kinetics and metabolism in laboratory animals and humans

Only few data are available on the kinetics of metallic or ionic palladium.

Palladium(II) chloride ($PdCl_2$) was poorly absorbed from the digestive tract (<0.5% of the initial oral dose in adult rats or about 5% in suckling rats after 3–4 days). Absorption/retention in adult rats was higher following intratracheal or intravenous exposure, resulting in total body burdens of 5% or 20%, respectively, of the dose administered, 40 days after dosing. Absorption after topical application was observed but not quantified.

After intravenous administration of different palladium compounds, palladium was detected in several tissues of rats, rabbits or dogs. The highest concentrations were found in kidney, liver, spleen, lymph nodes, adrenal gland, lung and bone. For example, 8–21% of the administered dose of palladium(II) chloride or sodium tetrachloropalladate(II) (Na_2PdCl_4) has been found in the liver or kidney of rats 1 day after dosing. After a 4-week dietary administration of palladium(II) oxide (PdO), measurable levels have been found only in the kidney of rats.

Only scarce data are available on the distribution of palladium from dental restorations in human tissues or fluids (e.g., in serum and saliva: about 1 µg/litre).

Transfer of small amounts of palladium to offspring via placenta and milk was seen with single intravenous doses of palladium(II) chloride in rats.

Information on the elimination and excretion of palladium is scarce and refers mostly to palladium(II) chloride and sodium tetrachloropalladate(II), which were found to be eliminated in faeces and urine. Urinary excretion rates of intravenously dosed rats and rabbits ranged from 6.4 to 76% of the administered dose during 3 h to 7 days. Elimination of palladium in faeces ranged in these studies from traces to 13% of the administered dose. Following oral administration of palladium(II) chloride, >95% of palladium was eliminated in faeces of rats due to non-absorption. Subcutaneous or topical treatment with palladium(II) sulfate ($PdSO_4$) or other palladium compounds resulted in detectable concentrations of palladium in the urine of guinea-pigs and rabbits.

Half-lives calculated for the elimination of palladium from rats (whole body, liver, kidney) ranged from 5 to 12 days.

Mean retention values determined at three time intervals (3 h, 24 h, 48 h) in rats injected intravenously with $^{103}PdCl_2$ showed little change with time for kidney, spleen, muscle, pancreas, thymus, brain and bone. They decreased slightly in liver and markedly in lung, adrenal gland and blood.

Owing to the ability of palladium ions to form complexes, they bind to amino acids (e.g., L-cysteine, L-cystine, L-methionine), proteins (e.g., casein, silk fibroin, many enzymes), DNA or other macromolecules (e.g., vitamin B_6).

The affinity of palladium compounds for nucleic acids was confirmed in many studies. *In vitro* experiments with palladium(II) chloride and calf thymus DNA indicated that palladium(II) interacts with both the phosphate groups and bases of DNA. Several palladium–organic complexes were observed to form bonds with calf thymus DNA or *Escherichia coli* plasmid DNA. Most of the complexes appear to interact via non-covalent binding, mainly via hydrogen bonding; in a few cases, however, indications for covalent binding were seen.

1.6 Effects on laboratory mammals and *in vitro* test systems

LD_{50} values for palladium compounds ranged, depending on compound and route tested, from 3 to >4900 mg/kg body weight, the most toxic compound being palladium(II) chloride, the least toxic, palladium(II) oxide. Oral administration caused the least toxicity. There were very similar intravenous LD_{50} values for palladium(II) chloride, potassium tetrachloropalladate(II) (K_2PdCl_4) and ammonium tetrachloropalladate(II) (($NH_4)_2PdCl_4$). Marked differences among the different routes of administration were demonstrated with palladium(II) chloride, showing in Charles-River CD1 rats LD_{50} values of 5 mg/kg body weight for the intravenous, 6 mg/kg body weight for the intratracheal, 70 mg/kg body weight for the intraperitoneal and 200 mg/kg body weight for the oral route. A higher oral LD_{50} value has been found in Sprague-Dawley rats.

Signs of acute toxicity of several palladium salts in rats or rabbits included death, decrease in feed and water uptake, emaciation, cases of ataxia and tiptoe gait, clonic and tonic convulsions, cardiovascular effects, peritonitis or biochemical changes (e.g., changes in activity of hepatic enzymes, proteinuria or ketonuria). Functional or histological changes in the kidney were found both with palladium compounds and with elemental palladium powder. There were also haemorrhages of lungs and small intestine.

Effects recorded in rodents and rabbits after short-term exposure to various palladium compounds refer mainly to changes in biochemical parameters (e.g., decrease in activity of hepatic microsomal enzymes or yield of microsomal protein). Clinical signs were sluggishness, weight loss, haematoma or exudations. Changes in absolute and relative organ weights and anaemia also occurred. One compound (sodium tetrachloropalladate(II) complexed with egg albumin) caused deaths in mice. Effective concentrations were in the milligram per kilogram body weight range. Histopathological effects have been observed in liver, kidney, spleen or gastric mucosa of rats 28 days after daily oral administration of 15 or 150 mg tetraammine palladium hydrogen carbonate ($[Pd(NH_3)_4](HCO_3)_2$)/kg body weight. Additionally, an increase in absolute brain and ovary weights at the 1.5 and 15 mg/kg body weight doses has been found.

The contribution of palladium to effects observed after single or short-term administration of palladium-containing dental alloy material is not clear.

There are also only scarce data available on effects from long-term exposure to palladium species (forms).

Mice given palladium(II) chloride (5 mg palladium/litre) in drinking-water from weaning until natural death showed suppression of body weight, a longer life span (in males, but not in females), an increase in amyloidosis of several inner organs and an approximate doubling of malignant tumours (see below).

Inhalative exposure of rats to chloropalladosamine (($NH_3)_2PdCl_2$) for about half a year caused slight, reversible (at 5.4 mg/m^3) or significant permanent (at 18 mg/m^3) changes in several blood serum and urine parameters, indicating damage mainly to liver and kidney (in addition to reduced body weight gain, changes in organ weights and glomerulonephritis). Adverse effects were also observed with enteral exposures, the no-observed-adverse-effect level being given as 0.08 mg/kg body weight.

Six months after a single intratracheal application of palladium dust (143 mg/kg body weight), several histopathological signs of inflammation were noted in the lungs of rats. Daily oral administration of palladium dust (50 mg/kg body weight) over 6 months resulted in changes in several blood serum and urine parameters of rats.

Skin tests of a series of palladium compounds in rabbits showed dermal reactions of different severity, resulting in the following ranking order: $(NH_4)_2PdCl_6$ > $(NH_4)_2PdCl_4$ > $(C_3H_5PdCl)_2$ > K_2PdCl_6 > K_2PdCl_4 > $PdCl_2$ > $(NH_3)_2PdCl_2$ > PdO. The first three compounds caused erythema, oedema or eschar in intact and abraded skin, the next three substances elicited erythema in abraded skin and the last two were not irritant. Palladium hydrochloride (formula not specified) also caused dermatitis in the skin of rabbits.

Eye irritation was observed with palladium(II) chloride and tetra-ammine palladium hydrogen carbonate (but not with palladium(II) oxide), both deposited on the eye surface of rabbits. Inhalation exposure to chloropalladosamine (≥ 50 mg/m^3) affected the mucous membranes of the eyes of rats (conjunctivitis, keratoconjunctivitis).

Some palladium compounds have been found to be potent sensitizers of the skin (palladium(II) chloride, tetraammine palladium hydrogen carbonate, palladium hydrochloride [formula not specified], palladium–albumin complexes). Palladium(II) chloride was a stronger sensitizer than nickel sulfate (NiSO$_4$) in the guinea-pig maximization test. Guinea-pigs induced with chromate, cobalt or nickel salts did not react after challenge with palladium(II) chloride. However, if induced with palladium(II) chloride, they reacted to nickel sulfate. Somewhat divergent results have been obtained in tests studying cross-reactivity between palladium and nickel by repeated open applications to the skin of guinea-pigs. In these experiments, animals were induced with palladium(II) chloride ($n = 27$) or nickel sulfate ($n = 30$) according to the guinea-pig maximization test method and then treated once daily for 10 days according to repeated open applications testing by applying the sensitizing allergen (palladium(II) chloride or nickel sulfate) as well as the possibly cross-reactive compound (nickel sulfate or palladium(II) chloride) and the vehicle topically in guinea-pigs. In this study, it remained unclear whether reactivity to palladium(II) chloride in animals sensitized with nickel sulfate was due to cross-reactivity or to the induction of sensitivity by the repeated treatments. On the other hand, reactivity to nickel sulfate in animals sensitized with palladium(II) chloride could be attributed to cross-reactivity. Respiratory sensitization (bronchospasms) has been observed in cats after intravenous administration of several complex palladium compounds. It was accompanied by an increase in serum histamine. Significant immune responses have been obtained with palladium(II) chloride and/or chloropalladates using the popliteal and auricular lymph node assay in BALB/c mice. Preliminary data in an animal model suggest that palladium(II) compounds may be involved in induction of an autoimmune disease.

There are insufficient data on the reproductive and developmental effects of palladium and its compounds. In one screening study, reduced testis weights were reported in mice that had received 30 daily subcutaneous doses of palladium(II) chloride at a total dose of 3.5 mg/kg body weight.

Palladium compounds may interact with isolated DNA *in vitro*. However, with one exception, mutagenicity tests of several palladium compounds with bacterial or mammalian cells *in vitro* (Ames test: *Salmonella typhimurium*; SOS chromotest: *Escherichia coli*; micronucleus test: human lymphocytes) gave negative results. Also, an *in vivo* genotoxicity test (micronucleus test in mouse) with tetraammine palladium hydrogen carbonate gave negative results.

Tumours associated with palladium exposure have been reported in two studies. Mice given palladium(II) chloride (5 mg Pd^{2+}/litre) in drinking-water from weaning until natural death developed malignant tumours, mainly lymphoma–leukaemia types and adenocarcinoma of the lung, at a statistically significant rate, but concomitant with an increased longevity in males, which may explain at least in part the increased tumour rate. Tumours were found at the implantation site in 7 of 14 rats (it was not clear whether the tumours were due to the chronic physical stimulus or to the chemical components) 504 days after subcutaneous implantation of a silver–palladium–gold alloy. No carcinogenicity study with inhalation exposure was available.

Palladium ions are capable of inhibiting most major cellular functions, as seen *in vivo* and *in vitro*. DNA/RNA biosynthesis seems to be the most sensitive target. An EC_{50} value of palladium(II) chloride for inhibition of DNA synthesis *in vitro* with mouse fibroblasts was

300 μmol/litre (32 mg Pd^{2+}/litre). Inhibition of DNA synthesis *in vivo* (in spleen, liver, kidney and testes) occurred in rats administered a single intraperitoneal dose of 14 μmol palladium(II) nitrate $(Pd(NO_3)_2)$/ kg body weight (1.5 mg Pd^{2+}/kg body weight).

Palladium applied in its metallic form showed no or little *in vitro* cytotoxicity, as evaluated microscopically.

A series of isolated enzymes having key metabolic functions have been found to be inhibited by simple and complex palladium salts. The strongest inhibition (K_i value for palladium(II) chloride = 0.16 μmol/ litre) was found for creatine kinase, an important enzyme of energy metabolism.

Many palladium–organic complexes have an antineoplastic potential similar to that of *cis*-dichloro-2,6-diaminopyridine-platinum(II) (*cis*-platinum, an anticancer drug).

The mode of action of palladium ions and of elemental palladium is not fully clear. Complex formation of palladium ions with cellular components probably plays a basic role initially. Oxidation processes may also be involved, due to the different oxidation states of palladium.

1.7 Effects on humans

There is no information on the effects of palladium emitted from automobile catalytic converters on the general population. Effects have been reported due to iatrogenic and other exposures.

Most of the case reports refer to palladium sensitivity associated with exposure to palladium-containing dental restorations, symptoms being contact dermatitis, stomatitis or mucositis and oral lichen planus. Patients with positive palladium(II) chloride patch tests did not necessarily react to metallic palladium. Only a few persons who showed positive patch test results with palladium(II) chloride showed clinical symptoms in the oral mucosa as a result of exposure to palladium-containing alloys. In one study, slight but non-significant changes in serum immunoglobulins were seen after placement of a silver–palladium alloy dental restoration.

Side-effects noted from other medical or experimental uses of palladium preparations include fever, haemolysis, discoloration or necrosis at injection sites after subcutaneous injections and erythema and oedema following topical application.

A few case reports reported skin disorders in patients who had exposure to palladium-containing jewellery or unspecified sources.

Serial patch tests with palladium(II) chloride indicated a high frequency of palladium sensitivity in special groups under study. Several recent and large-sized studies from different countries found frequencies of palladium sensitivity of 7–8% in patients of dermatology clinics as well as in schoolchildren, with a preponderance in females and younger persons. Compared with other allergens (about 25 were studied), palladium belongs to the seven most frequently reacting sensitizers (ranked second after nickel within metals). Solitary palladium reactions (monoallergy) occurred with a low frequency. Mostly, combined reactions with other metals (multisensitivity), primarily nickel, have been observed.

To date, the most often identified sources of palladium sensitization for the general population are dental restorations and jewellery.

There are few data on adverse health effects due to occupational exposure to palladium. Few PGM workers (2/307; 3/22) showed positive reactions to a complex palladium halide salt in sensitization tests (skin prick test; radioallergosorbent test; monkey passive cutaneous anaphylaxis test). Some workers (4/130) of an automobile catalyst plant had positive reactions in prick tests with palladium(II) chloride. A review article (without details) reported on a frequent occurrence of allergic diseases of the respiratory passages, dermatoses and affections of the eyes among Russian PGM production workers. Single cases of allergic contact dermatitis have been documented for two chemists and a metal worker. A single case of palladium salt-induced occupational asthma has been observed in the electronics industry.

Subpopulations at special risk of palladium allergy include people with known nickel allergy.

1.8 Effects on other organisms in the laboratory and field

Several palladium compounds have been found to have antiviral, antibacterial and/or fungicidal properties. Standard microbial toxicity tests under environmentally relevant conditions have rarely been conducted. A 3-h EC_{50} of 35 mg/litre (12.25 mg palladium/litre) has been obtained for the inhibitory effect of tetraammine palladium hydrogen carbonate on the respiration of activated sewage sludge.

Those palladium compounds that have been tested for effects on aquatic organisms have been found to be of significant toxicity. Two palladium complexes (potassium tetrachloropalladate(II) and chloro-palladosamine) present in nutrient solution caused necrosis at 2.5–10 mg palladium/litre in the water hyacinth (*Eichhornia crassipes*). The acute toxicity (96-h LC_{50}) of palladium(II) chloride to the freshwater tubificid worm *Tubifex tubifex* was 0.09 mg palladium/litre. A minimum 24-h lethal concentration of 7 mg palladium(II) chloride/li-tre (4.2 mg palladium/litre) has been reported for the freshwater fish medaka (*Oryzias latipes*). In all cases, palladium compounds had a toxicity similar to that of platinum compounds.

Toxicity tests on aquatic organisms conducted according to Organisation for Economic Co-operation and Development guidelines have been performed only for tetraammine palladium hydrogen carbonate. They resulted in a 72-h EC_{50} value of 0.066 mg/litre (corres-ponding to 0.02 mg palladium/litre) (cell multiplication inhibition test with *Scenedesmus subspicatus*), a 48-h EC_{50} of 0.22 mg/litre (0.08 mg palladium/litre) (immobilization of *Daphnia magna*) and a 96-h LC_{50} of 0.53 mg/litre (0.19 mg palladium/litre) (acute toxicity to rainbow trout *Oncorhynchus mykiss*). The no-observed-effect concentrations (NOECs) were given as 0.04 mg/litre (0.014 mg palladium/litre) (algae), 0.10 mg/litre (0.05 mg palladium/litre) (*Daphnia magna*) and 0.32 mg/litre (0.11 mg palladium/litre) (fish). All these values have been based on nominal concentrations. However, corresponding measured concentrations have often been found to be much lower and variable, the reasons for this being unclear. For the immobilization test with *Daphnia magna*, values based on the time-weighted mean mea-sured concentrations have been calculated, resulting in a 48-h EC_{50} of 0.13 mg/litre (0.05 mg palladium/litre) and a NOEC of 0.06 mg/litre (0.02 mg palladium/litre). Phytotoxic effects have also been observed in terrestrial plants after addition of palladium(II) chloride to the nutrient solution. They include inhibition of transpiration at 3 mg/litre (1.8 mg palladium/litre), histological changes at 10 mg/litre (6 mg palladium/litre) or death at 100 mg/litre (60 mg palladium/litre) in Kentucky bluegrass (*Poa pratensis*). Dose-dependent growth retardation and stunting of the roots occurred in several crop plants, the most sensitive being oats, affected at about 0.22 mg palladium(II) chloride/litre (0.132 mg palladium/litre).

No information has been located in the literature on the effects of palladium on terrestrial invertebrates or vertebrates.

There are no field observations available.

2. IDENTITY, PHYSICAL AND CHEMICAL PROPERTIES AND ANALYTICAL METHODS

2.1 Identity

Palladium (Pd) belongs to the platinum group metals (PGMs), which comprise six closely related metals: platinum (Pt), palladium, rhodium (Rh), ruthenium (Ru), iridium (Ir) and osmium (Os). These metals commonly occur together in nature and are among the scarcest of the metallic elements. Along with gold (Au) and silver (Ag), they are known as precious and noble metals. Palladium is a steel-white metal, does not tarnish in air and has the lowest density and lowest melting point of the PGMs. The most important palladium compounds are listed in Table 1.

2.2 Physical and chemical properties

2.2.1 Palladium metal

Purified metals between 99.9% and 99.999% palladium are available for chemical or medical use as foil, granules, powder, rod or wire (Aldrich, 1996). Table 2 lists atomic and crystal data as well as physical properties of palladium metal.

Palladium metal resists oxidation at ordinary temperatures. Palladium has a strong catalytic activity, especially for hydrogenation and oxidation reactions.

The reaction of palladium powder with oxygen may cause a fire hazard. This is particularly the case in the presence of combustible substances (e.g., carbon catalysts). On contact with palladium powder, hydrogen peroxide and other peroxides, concentrated formic acid and hydrazine are expected to decompose rapidly (Degussa, 1995).

2.2.2 Palladium compounds

Several hundred palladium compounds in various oxidation states (Table 3) are known from the scientific literature, but only a few of them are of economic relevance (see also section 3.2.4). In its

Table 1. Chemical names, synonyms and formulas of
selected palladium compounds[a]

Chemical name	Synonyms	Molecular formula	CAS registry no.
Palladium		Pd	7440-05-3
Ammonium hexachloro-palladate(IV)		$(NH_4)_2PdCl_6$	19168-23-1
Ammonium tetrachloro-palladate(II)		$(NH_4)_2PdCl_4$	13820-40-1
Bis(1,5-diphenyl-1,4-pentadien-3-one) palladium(0)	Bis(dibenzylidene-acetone) palladium	$Pd(C_{17}H_{14}O)_2$	32005-36-0
Bis(2,4-pentane-dionato) palladium(II)	Bis(acetylacetonato) palladium(II)	$Pd(C_5H_7O_2)_2$	14024-61-4
cis-Diammine-dichloro-palladium(II)	Chloropalladosamine	$(NH_3)_2PdCl_2$	14323-43-4
trans-Diammine-dichloro-palladium(II)		$(NH_3)_2PdCl_2$	13782-33-7
Diammine-dinitro-palladium(II)		$Pd(NO_3)_2(NH_3)_2$	not available
trans-Dichloro-bis-(triphenyl-phosphine) palladium(II)		$[(C_6H_5)_3P]_2PdCl_2$	13965-03-2
Dichloro(1,5-cyclooctadiene) palladium(II)		$PdCl_2(C_8H_{12})$	12107-56-1
Hydrogen tetrachloro-palladate(II)	Tetrachloropalladous acid	H_2PdCl_4	16970-55-1
Palladium(II) acetate	Palladium diacetate	$Pd(CH_3COO)_2$	3375-31-3
Palladium(II) chloride	Palladous chloride Palladium dichloride	$PdCl_2$	7647-10-1

Table 1 (contd).

Chemical name	Synonyms	Molecular formula	CAS registry no.
Palladium(II) iodide	Palladous iodide	PdI_2	7790-38-7
Palladium(II) nitrate	Palladous nitrate	$Pd(NO_3)_2$	10102-05-3
Palladium(II) oxide	Palladium monoxide	PdO	1314-08-5
Palladium(II) sulfate	Palladous sulfate	$PdSO_4$	13566-03-5
Potassium hexachloro-palladate(IV)		K_2PdCl_6	16919-73-6
Potassium tetrachloro-palladate(II)	Potassium palladium chloride	K_2PdCl_4	10025-98-6
Sodium tetrachloro-palladate(II)		$Na_2PdCl_4 \cdot 3H_2O$	13820-53-6
Tetraammine palladium(II) chloride	Tetraammine palladium(II) dichloride	$[Pd(NH_3)_4]Cl_2$	13815-17-3
Tetraammine palladium hydrogen carbonate	TPdHC Tetramminepalladium hydrogen carbonate	$[Pd(NH_3)_4](HCO_3)_2$	134620-00-1
Tetrakis(tri-phenyl-phosphine) palladium(0)		$Pd[(C_6H_5)_3P]_4$	14221-01-3

[a]   Compiled from Degussa (1995); Aldrich (1996); Kroschwitz (1996); Lovell, personal communication, Johnson Matthey plc, August 1999.

compounds, palladium most commonly exhibits an oxidation state of +2. Compounds of palladium(IV) are fewer and less stable. Like the other PGMs, palladium has a strong disposition to form coordination complexes. The complexes are predominantly square planar in form. In addition, palladium forms a series of organic complexes, reviewed in Kroschwitz (1996). The organometallic palladium(II) compounds include σ-bound alkyls, aryls, acyls and acetylides as well as π-bound (di)olefins, alkyls and cyclopentadienyls.

Table 2. Atomic and physical properties of palladium metal[a,b]

Property	Palladium
Classification	Transition metal
Standard state	Solid
Specimen	Available as foil, granules, powder, rod, shot, sponge or wire
Atomic number	46
Relative atomic mass	106.42
Abundance of major natural isotopes[c]	105 (22.3%),106 (27.3%), 108 (26.5%)
Colour/form	Steel-white, ductile metal
Odour	Odourless
Electronegativity (Pauling scale)	2.2
Crystal structure	Cubic
Atomic radius (nm)	0.179
Melting point (°C)	1554
Boiling point (°C)	2940
Exposure to heat or flame	Non-combustible; no decomposition
Density at 20 °C (g/cm^3)	12.02
Reduction potential Pd/Pd^{2+} of aqua complexes	+0.92[d] (at pH 1)
Solubility[e,f]	Insoluble in water (pH 5–7), acetic acid (99%), hydrofluoric acid (40%), sulfuric acid (96%) or hydrochloric acid (36%) at room temperature Slightly soluble in sulfuric acid (96%; 100 °C) and sodium hypochlorite solution (20 °C) Soluble in aqua regia (3:1 HCl/HNO$_3$ at 20 °C) and nitric acid (65%; 20 °C)

[a] Information valid for ^{106}Pd unless otherwise noted.
[b] Compiled from Smith et al. (1978); Lide (1992); Budavari et al. (1996).
[c] ^{103}Pd is not a naturally occurring isotope.
[d] Holleman & Wiberg (1995).
[e] Degussa (1995).
[f] For solubility in biological media, see section 6.1.

Physical and chemical properties of selected palladium compounds are given in Table 4.

Table 3. Examples of important palladium compounds by oxidation state[a]

Oxidation state	Electronic configuration	Examples
Pd(0)	d10	Pd, Pd[(C$_6$H$_5$)$_3$P]$_4$, Pd(PF$_3$)$_4$
Pd(II)	d8	[Pd(OH$_2$)$_4$]$^{2+}$(aq), [Pd(NH$_3$)$_4$]$^{2+}$, [Pd(NH$_3$)$_2$Cl$_2$], PdF$_2$, PdCl$_2$, etc., PdO, [PdCl$_4$]$^{2-}$, [PdSCN$_4$]$^{2-}$, [PdCN$_4$]$^{2-}$, [Pd$_2$Cl$_6$]$^{2-}$, salts, complexes
Pd(IV)	d6	PdO$_2$, PdF$_4$, [PdCl$_6$]$^{2-}$

[a] Compiled from Cotton & Wilkinson (1982).

2.3 Analytical methods

Palladium (as a solution of palladium(II) nitrate in the mg/litre concentration range) is frequently used as a chemical modifier to overcome interferences with the determination of various trace elements in biological materials by graphite furnace atomic absorption spectrometry (GF-AAS) (Schlemmer & Welz, 1986; Taylor et al., 1998). Care must be taken, therefore, in analytical laboratories using palladium chemical modifiers to avoid contamination when measuring palladium by the GF-AAS technique.

2.3.1 Sample collection and pretreatment

Palladium is rarely found in significant concentrations in any kind of environmental material. Environmental and biological materials being investigated for very low levels of palladium need to be sampled in large amounts, with possible difficulty in homogenization, digestion, storage and matrix effects. In order to obtain enough of the analyte for accurate determinations and to separate the palladium from the sample matrix and interfering elements, preconcentration is often necessary.

Several chemical methods for the separation and preconcentration of palladium have been developed — for example, extraction with various agents, separation with ion-exchange resins, co-precipitation with tellurium or mercury and fire assay (Eller et al., 1989; Tripkovic et al., 1994). For example, palladium(II) in aqueous solution can be extracted by diethyldithiocarbamate (Shah & Wai, 1985; Begerow et al., 1997a), N-p-methoxyphenyl-2-furylacrylohydroxamic acid (Abbasi, 1987) or 1-decyl-N,N'-diphenylisothiouronium bromide (Jones et al., 1977).

Table 4. Physical and chemical properties of selected palladium compounds

Chemical name	Appearance	Molecular mass (g)	% Pd	Melting point (°C)[a]	Solubility in water	Solubility in other solvents	Relative density (g/cm³)	Reference
Bis(acetylacetonato) palladium(II)	yellow crystals	304.64	34.9					NAS (1977)
Bis(dibenzylidene-acetone) palladium(0)	purple powder	575.02	18.5					NAS (1977)
Diamminedinitro-palladium(II)	yellow	232.5	45.8		slightly soluble	soluble in ammonium hydroxide		Degussa (1995)
Dichloro(1,5-cyclo-octadiene) palladium(II)	yellow crystals	285.51	37.3					NAS (1977)
Palladium(II) chloride	rust colour powder	177.33	60	675 or 501[b] (dec.)	soluble	soluble in hydrochloric acid, alcohol, acetone	4	Sax & Lewis (1987); Budavari et al. (1996)
Palladium(II) acetate	reddish-brown crystals	224.51	47.4	200 (dec.)	insoluble	soluble in hydrochloric acid or potassium iodide solution		Sax & Lewis (1987); Budavari et al. (1996)
Palladium(II) iodide	black powder	360.21	29.5	350 (dec.)	insoluble	soluble in potassium iodide solution	6	Sax (1979); Lewis (1987)
Palladium(II) oxide	black-green or amber solid	122.4	87	750 (dec.)	insoluble	soluble in dilute aqua regia, 48% hydrobromic acid	8.3	Sax (1979); Sax & Lewis (1987)

Table 4 (contd).

Chemical name	Appearance	Molecular mass (g)	% Pd	Melting point (°C)[a]	Solubility in water	Solubility in other solvents	Relative density (g/cm³)	Reference
Palladium(II) acetate trimer	gold brown crystals	673.53	47.4		insoluble	soluble in acetic acid		NAS (1977)
Palladium(II) nitrate	brown salt	229.94 (anhydrous)	~46.2	dec.	soluble	soluble in dilute nitric acid		Sax & Lewis (1987); Budavari et al. (1996)
Potassium chloropalladate	cubic red crystals	397.3	53.6	(dec.)			2.7	Sax (1979)
Potassium tetrachloropalladate(II)	reddish-brown crystals	326.4	32.6	524	soluble	slightly soluble in hot alcohol	2.7	Sax & Lewis (1987)
Sodium tetrachloropalladate(II)	red brown powder	294.21	37					NAS (1977)
Tetraammine-palladium(II) chloride	yellow	245.4	43.4		soluble			Degussa (1995)
Tetraammine palladium hydrogen carbonate		219.4	48.5	181 (dec.)	soluble (56.2 g/litre at 20 °C)			Johnson Matthey (2000)
Tetrachloropalladic(II) acid	dark brown	250.2	42.5			only stable in solution of hydrochloric acid		

Table 4 (contd).

Chemical name	Appearance	Molecular mass (g)	% Pd	Melting point (°C)[a]	Solubility in water	Solubility in other solvents	Relative density (g/cm³)	Reference
Tetrakis(triphenyl-phosphine) palladium(0)	yellow crystals	1155.58	9.2		insoluble	soluble in acetone, chlorinated hydrocarbons, benzene		NAS (1977)
trans-Diamminedichloro-palladium(II)	orange crystals	211.39	50.3		soluble (2.7 g/litre)	soluble in ammonium hydroxide		NAS (1977)
trans-Dichlorobis-(triphenylphosphine) palladium(II)	yellow crystals	701.91	15.2					NAS (1977)

[a] dec. = decomposes.
[b] From Sax (1979).

Cellulose ion exchangers (Kenawy et al., 1987), 2,2'-dipyridyl-3-(4-amino-5-mercapto)-1,2,4-triazolylhydrazone supported on silica gel (Samara & Kouimtzis, 1987) or automated on-line column separation systems (Schuster & Schwarzer, 1996), were used to preconcentrate traces of palladium(II) from water samples.

For laboratories engaged in analyses of geological samples, the fire assay fusion seems to be the preferred method of dissolving and concentrating palladium. Palladium metal can be preconcentrated using either a lead collection or a nickel sulfide collection. The sensitivity of the nickel sulfide fire assay is limited by background palladium introduced by the high amounts of chemicals (e.g., nickel) employed (McDonald et al., 1994).

With biological materials, homogeneous sampling is difficult and often requires destructive methods, resulting in the loss of all information about the palladium species. In many of the analytical procedures, samples have been ashed to destroy organic materials and then treated with strong acids to yield solutions for palladium determination. Only the total content of palladium and its isotopes can then be determined. For the analysis of palladium in urine, the untreated original sample is usually unsuitable. Freeze-drying or a wet ashing procedure with subsequent reduction of volume is necessary for most analytical methods. For complex matrices such as blood, removal of the organic sample matrix combined with dilution to reduce the content of total dissolved solids is recommended to avoid blockages of the sampling cone and signal instabilities when using inductively coupled plasma mass spectrometry (ICP-MS). Strong mineral acids are most frequently applied for matrix decomposition. For blood, serum and urine digestion, ultraviolet (UV) photolysis has also been found to be useful.

2.3.2 Reference materials

The availability of certified reference materials is of great value for laboratories engaged in analytical chemistry. For palladium analysis, there are only few international standard reference materials, which are directly traceable to the Standard Reference Material (SRM) of the US National Bureau of Standards (NBS). Single-element AAS standards are offered at 1 mg/ml — for example, by Aldrich (1996) — or can be prepared according to APHA et al. (1989). To our knowledge, interlaboratory comparison programmes for the determination of environmental palladium are not yet available.

2.3.3   Analysis

Analytical methods are summarized in Table 5. Current measurement techniques do not allow separate species of palladium (metal or palladium(II) compounds) to be differentiated when more than one form is present. Almost all measurements of palladium in environmental and other samples to date have been for total palladium.

In analytical laboratories, physical methods have widely replaced wet chemical and colorimetric analytical methods for reasons of economy and speed. Methods such as neutron activation analysis, total reflection X-ray fluorescence analysis and, above all, ICP-MS and GF-AAS are used after appropriate enrichment procedures. If palladium is brought into solution by appropriate separation methods, all PGMs can be determined in the presence of each other by X-ray fluorescence or ICP analysis, for example.

Using ICP-MS, it is possible to detect palladium in urine or blood samples of persons without occupational exposure, whereas the detection limits of AAS methods are higher by a factor of about 3 or more (see Table 5).

Table 5. Analytical methods for palladium determination

Matrix/medium	Sample treatment (decomposition/separation)	Determination method[a]	Limit of detection[b]	Comments[c]	References
Air					
Particulate matter	air filtration through Teflon membrane	XRF	0.001 µg/m^3 ^{d}		Lu et al. (1994)
Particulate matter	air filtration through Teflon membrane	XRF	0.0005 µg/m^3 ^{d}		Gertler (1994)
Car exhaust					
Particulate matter from exhaust pipe of cars	bubbling through nitric acid absorbent solution and filtering through cellulose ester filter; mineralization: acid-assisted microwave digestion	quadrupole ICP-MS	3.3 ng/litre	mathematical corrections for spectral interferences	Moldovan et al. (1999); Gomez et al. (2000)
Water					
Aqueous solution	extraction with 1-decyl-N,N'-diphenyl-isothiouronium bromide in variety of organic liquids	AAS	<0.1 mg/litre	co-extraction of noble metals	Jones et al. (1977)
Aqueous solution	extracted by bismuth diethyldithio-carbamate into chloroform at pH 3.5	NAA	0.4 ± 0.1 ng/litre	Pd(II); 5 litres of river water were extracted	Shah & Wai (1985)
Water samples	2,2'-dipyridyl-3-(4-amino-5-mercapto)-1,2,4-triazolylhydrazone, supported on silica gel column	AAS	4 µg/litre	Pd(II); samples were preconcentrated by a factor of 100	Samara & Kouimtzis (1987)

Table 5 (contd).

Matrix/medium	Sample treatment (decomposition/separation)	Determination method[a]	Limit of detection[b]	Comments[c]	References
Spring water sample	adsorption on sulfonated dithizone-loaded resin; direct introduction into the furnace	GF-AAS	22 ± 2 ng/litre		Chikuma et al. (1991)
Pure waters	acidification and adsorption onto activated charcoal; palladium is redissolved with aqua regia following ashing of the charcoal	ICP-MS	0.3–0.8 ng/litre	1-litre sample volume; preconcentration factor of 200	Hall & Pelchat (1993)
Groundwater	only filtration and acidification with nitric acid	ICP-MS	5 ng/litre	convenient for determination of multimetals; strontium interferes with ^{105}Pd	Stetzenbach et al. (1994)
Aqueous solution	preconcentration in a microcolumn loaded with N,N-diethyl-N'-benzoyl-thiourea	GF-AAS	13–51 ng/litre	analysis of ethanol eluate	Schuster & Schwarzer (1996)
Geological materials					
Rock, water	extraction with selenium via a co-precipitation technique	ZAAS	1–3 ng/ml		Eller et al. (1989)
Various geological materials	extraction with aqua regia/hydrofluoric acid, co-precipitation with tellurium	AAS	0.031 µg/ml analyte solution	determination in solution using an argon-stabilized arc	Tripkovic et al. (1994)
Various geological materials	fire assay (nickel sulfide)	GF-AAS	2 µg/kg		Zereini (1996)

Table 5 (contd).

Matrix/medium	Sample treatment (decomposition/separation)	Determination method[a]	Limit of detection[b]	Comments[c]	References
Soil and dust					
Roadside dust	dissolution by wet ashing: isolation by anion exchange	FAAS	15 µg/kg[d]		Hodge & Stallard (1986)
Roadside dust, soil	fire assay (nickel sulfide); digestion with hydrochloric acid	flameless AAS	2 µg/kg	appropriate for geological materials	Zereini et al. (1993)
Human blood and urine					
Blood	wet ashing with nitric acid/perchloric acid; extraction with tri-n-octylamine from hydrochloric acid solution	flameless AAS	0.4 µg/litre	determinations on spiked samples; quantity 15 ml	Tillery & Johnson (1975)
Blood, urine	urine: evaporation of urine samples blood: wet ashing with nitric acid/ perchloric acid; extraction with tri-n-octylamine from hydrochloric acid; aspiration into air–acetylene flame	AAS	blood: 0.9 µg/100 ml; urine: 0.3 µg/litre		Johnson et al. (1975a,b)
Whole blood, urine	decomposition with nitric acid/ perchloric acid	flameless AAS	0.01 µg/g blood; 0.003 µg/g urine	rapid method (5-g blood sample) (50-g urine sample)	Jones (1976)
Urine	direct measurement	ICP-AES	16 ng/ml	5-µl samples	Matusiewicz & Barnes (1988)

Table 5 (contd).

Matrix/medium	Sample treatment (decomposition/separation)	Determination method[a]	Limit of detection[b]	Comments[c]	References
Urine	adjustment to pH 4, conversion to the pyrrolidinedithiocarbamate complex, extraction into 4-methyl-2-pentanone	ETA-AAS	0.02 µg/litre		Begerow et al. (1997a)
Urine	acidification with nitric acid	quadrupole ICP-MS	0.03 µg/litre	no further sample treatment other than calibration	Schramel et al. (1997)
Whole blood, urine	samples mixed with hydrogen peroxide/nitric acid; digestion by UV photolysis	sector field ICP-MS	0.2 ng/litre	cleaning of all materials resulted in a drastic reduction of blanks	Begerow et al. (1997b,c)
Other biological materials					
Meat	dry ashing of homogenized meats (11–12 kg): decomposition with aqua regia/hydrofluoric acid	NAA	0.5 µg/kg meat	radionuclide [103]Pd	Koch & Roesmer (1962)
Human organ material, blood	wet mineralization with sulfuric acid/ nitric acid/hydrogen peroxide; extraction with diethylammonia-diethyldithiocarbamate in chloroform	ESA	2.5 µg/g	1 g organ material	Geldmacher-von Mallinckrodt & Pooth (1969)
Human hair, faeces	digestion with nitric acid/perchloric acid; aspiration into air–acetylene flame	AAS	20 ng/g (hair), 1 ng/g (faeces)		Johnson et al. (1975a,b)

27

Table 5 (contd).

Matrix/medium	Sample treatment (decomposition/separation)	Determination method[a]	Limit of detection[b]	Comments[c]	References
Rice, tea, human hair	digestion with perchloric acid/nitric acid; cathodic stripping voltammetric determination by mixed binder carbon paste electrode containing dimethyl-glyoxime	VD	0.1 µg/g	samples spiked with Pd^{2+} were examined; simultaneous determination of Hg, Co, Ni, Pd	Zhang et al. (1996)
Biological materials, fresh waters	extracted with N-p-methoxyphenyl-2-furylacrylohydroxamic acid in isoamyl alcohol at pH 2.7–3.5	spectro-photometry	0.1 µg/litre	enrichment of Pd(II) 15 times	Abbasi (1987)
Marine macrophytes	dry ashing and wet digestion; purification with an anion-exchange resin	GF-AAS	0.11 µg/kg[d]		Yang (1989)
Ash of plant tissue	ashing at 870 °C and digestion in hydrofluoric acid/aqua regia	ICP-MS	0.5–1 µg/kg		Rencz & Hall (1992)
Various foods	digestion with nitric acid; calibration with rhodium and rhenium as internal standards	ICP-MS	0.9 µg/kg (peanut oil), 0.1 µg/kg (water)	0.5-g samples	Zhou & Liu (1997)
Release from dental alloys					
Cell culture medium	centrifugation	FAAS	35 µg/litre	the supernate of the cell culture medium was analysed	Wataha et al. (1992)
Cell culture medium	direct measurement of cell culture extracts	FAAS	20 µg/litre		Wataha et al. (1995a)

28

Table 5 (contd).

Matrix/medium	Sample treatment (decomposition/separation)	Determination method[a]	Limit of detection[b]	Comments[c]	References
Artificial saliva	direct measurement of the solution	AAS	30 µg/litre		Pfeiffer & Schwickerath (1995)
Miscellaneous material					
Diverse samples	digestion in aqua regia; separation on anion-exchange and concentrator column; eluants: sodium perchlorate/ hydrochloric acid	UV-D	1 µg/litre		Rocklin (1984)
Catalytic converter block	catalytic converter sample leached in hydrochloric acid/sodium chloride for 12 h; separation of the chloride complexes by electrophoresis	CZE-UV	1.4 µg/ml	simultaneous determination of Pd^{2+} and Pt^{4+}	Baraj et al. (1996)

[a] AAS = atomic absorption spectrometry; CZE-UV = capillary zone electrophoresis, using direct UV absorbance detection; ESA = emission spectrochemical analysis; ETA-AAS = atomic absorption spectrometry with electrothermal atomization; FAAS = flame atomic absorption spectrometry; GF-AAS = graphite furnace atomic absorption spectrometry; ICP-AES = inductively coupled plasma atomic emission spectrometry; ICP-MS = inductively coupled plasma mass spectrometry; NAA = neutron activation analysis; UV-D = ultraviolet detection; VD = voltammetric determination; XRF = X-ray fluorescence analysis; ZAAS = Zeeman graphite furnace atomic absorption spectrometry.

[b] The limit of detection normally represents the concentration of analyte that will give a signal to noise ratio of 2.

[c] No information about the oxidation state is given, except it is stated that palladium(II) was determined.

[d] Lowest value indicated.

29

3. SOURCES OF HUMAN AND ENVIRONMENTAL EXPOSURE

3.1 Natural occurrence

PGMs occur naturally in very low concentrations ubiquitously in the environment (Table 6). The fraction of palladium within PGMs is approximately 20% (Renner & Schmuckler, 1991; Renner, 1992).

A concentration of palladium below 1 µg/kg in the upper continental crust is estimated. This is in accordance with a mean value of 0.4 µg palladium/kg proposed by Wedepohl (1995). Together with the other PGMs, palladium occurs at a concentration below 1 ng/kg in seawater.

3.2 Anthropogenic "sources" of palladium

3.2.1 Production levels and processes for palladium metal

Nearly all of the world's supply of PGMs is extracted from deposits in four countries: the Republic of South Africa, Russia, Canada, and the USA. The primary production of palladium for these and other producing countries is listed in Table 7.

The largest fraction of palladium is recovered as a by-product of copper and nickel sulfide ore refining (Russia and Canada) or as alloys of the PGMs from primary PGM deposits (South Africa, USA). PGMs are generally found in mixtures of varying proportions.

Because PGMs are very expensive to mine and purify, a high proportion of PGMs are recycled by the users or by the producers (e.g., catalysts) and do not appear on the market. Averaged over all the PGMs, the quantity recycled amounts at present to about 20% of primary production, with the greatest emphasis on platinum and palladium (Loebenstein, 1996). Therefore, supply figures essentially reflect only mined products and sales from the former Soviet Union/Russia.

Table 6. Distribution of all platinum group metals in the environment[a]

Region	Estimated concentration of PGMs
Earth	~30 mg/kg
Mantle (siliceous lithosphere)	~0.05 mg/kg
Earth's crust (attainable by mining)	~0.01 mg/kg
Hydrosphere	$<10^{-6}$ mg/litre
Biomass (dry matter)	$<10^{-7}$ mg/kg

[a] Adapted from Renner & Schmuckler (1991).

Table 7. Palladium output, by country[a]

Country	Palladium output (tonnes)							
	1981	1987	1990	1993	1995	1996	1997	1998
Soviet Union/ Russia[b]	44.5	55.7	58.2	71.5	40	37.9	149.4	180.5
South Africa	28.3	33.9	38.2	43.4	44	103.6	56.3	56.3
Canada and USA	5.0	5.9	11.5	11.5	13	7.5	17.0	20.5
Others	2.2	2.8	2.2	2.2	2.0	3.4	3.0	3.7
Total (tonnes)	80.0	98.3	110.1	128.6	99.0	152.4	225.7	261.0

[a] 1981–1993 from Kroschwitz (1996); 1995 from Loebenstein (1996); 1996 from MMAJ (1999); 1997–1998 from Cowley (1998, 1999).
[b] Russian values for 1981–1996: sales to the West.

3.2.1.1 Production processes

Several enrichment steps follow the mining of ores, either under-ground or by opencast (strip) mining (Renner & Tröbs, 1986; Renner, 1992; Kroschwitz, 1996). In the case of PGMs from South Africa, the crude ore is first crushed and pulverized. The metal sulfide particles are then separated from the gangue by froth flotation.

The PGM sulfide pulp is melted in an electric furnace to produce matte containing principally copper, nickel and iron sulfides, together with the PGMs. Nickel and copper are separated by acid leaching to yield the final PGM concentrate. The concentrate is then refined, using either a selective dissolution and precipitation technique or a solvent extraction process, whereby ammonium-chloro-complexes (e.g., chloropalladosamine) are formed. Thermal decomposition (calcination) of chloropalladosamine gives the impure palladium sponge.

3.2.1.2   Recycling

Processing of recycled material, including both new and old scrap, resulted in the recovery of an estimated 60 tonnes of PGMs during 1995 in the USA (Loebenstein, 1996).

Scrapped automobile catalysts contributed 5.4 tonnes to the world palladium demand of 72 tonnes in the automotive sector in 1998 (Cowley, 1999). According to Cowley (1999), 5.12 tonnes of palladium were recovered from old automobile catalysts in North America and Japan in 1998. The ultimate fate of many automobile catalysts may be disposal in waste dumps. Recycling methods comprise pyrometallurgical refining techniques that are similar to those described previously in section 3.2.1.1 (e.g., Degussa, 1997). The quantities recovered from scrapped electrical components (electrical contacts) are larger, but again no figures exist (Cowley, 1997). It is expected that some of the utilized palladium ends up in wastes or incineration ashes. It can also be expected that dental alloys are completely recycled if dental prostheses have to be replaced.

Recycling rates for electronic equipment and automobile catalysts depend on both economic and political decisions in the different countries.

3.2.2   Processes for the production of important palladium compounds

Many palladium compounds are in use as catalysts, as precursors of metallic palladium, in preparative chemical production, in photography, in electroplating and in medicine. The production processes of some important compounds are described below:

- ▶ *Ammine complexes of palladium*: Addition of ammonia to solutions of palladium(II) chloride first causes the formation of a pink precipitate of the binuclear complex $Pd(NH_3)_4PdCl_4$, Vauquelin's salt, which is converted to soluble tetraammine palladium(II) chloride by further addition of ammonia. Acidification of this solution yields the sparingly soluble light-yellow *trans-*

diamminedichloropalladium(II). Ammonium hexachloropalla-date(IV) is an oxidation product of ammonium tetrachloropalla-date(II).

▶ *Palladium(II) acetate*: This compound is prepared from palladium sponge (or nitrate) and glacial acetic acid.

▶ *Palladium(II) chloride*: Palladium(II) chloride is prepared by the careful evaporation of a solution of hydrogen tetrachloropalla-date(II) in hydrochloric acid, preferably in a rotary evaporator.

▶ *Palladium(II) nitrate*: This compound is prepared from palladium and nitric acid (Renner, 1992).

▶ *Palladium(II) oxide*: Palladium(II) oxide is obtained by reaction of palladium black (powder) with oxygen or air at 750 °C. Decomposition occurs at 850 °C. A catalytically active palladium preparation analogous to platinum(IV) oxide ($PtO_2[H_2O]_x$) can be obtained by evaporating a solution of hydrogen tetrachloropalla-date(II) and sodium nitrate and fusing the product (Cotton & Wilkinson, 1982; Neumüller, 1985).

▶ *Tetrachloropalladic(II) acid*: The metal is dissolved in hydro-chloric acid/chlorine or hydrochloric acid/nitric acid. If disso-lution occurs below about 50 °C, hexachloropalladic(IV) acid is formed first. Commercial solutions in hydrochloric acid contain 20% palladium (Renner, 1992).

3.2.3 Uses of palladium metal

From Table 8, it can be seen that demand for palladium, in particular for use in automobile catalysts, is increasing.

3.2.3.1 Electronics and electrical technology

Palladium metal or silver–palladium powder pastes are important products in the production of many electronic components. The metallization process is often carried out with silver–palladium thick film paste. The pastes are used in active components such as diodes, transistors, integrated circuits, hybrid circuits and semiconductor memories. They are also needed for passive electronic components,

Table 8. Western world palladium metal demand according to application[a]

Use	Demand (tonnes) (1993)[b]	% (1993)	Demand (tonnes) (1996)[c]	% (1996)	Demand (tonnes) (1998)[c]	% (1998)
Electrical equipment	61.1	45.0	64.1	31.9	64.4	24.8
Dental	37.6	27.7	40.7	20.3	38.3	14.7
Automotive emission control catalysts	20.3 (+3.2)	17.3	71.9 (+4.5)	38.1	139	53.4
Other (catalysts, jewellery, chemicals)	13.5	9.9	17.6	8.8	18.5	7.1
Western sales to China	–	–	1.9	0.9	–	–
Total	135.7	99.9	200.7			

[a] Comprises primary and refined secondary materials. Numbers in parentheses represent the quantity of palladium recovered from the automobile catalyst industry (internal cycle).
[b] Adapted from Kroschwitz (1996).
[c] Adapted from Cowley (1997, 1999).

such as very small multilayer ceramic capacitors, thick film resistors or conductors.

Silver–palladium alloys are used for electrical contacts, and other palladium alloys are used for electrical relays and switching systems in telecommunication equipment. In low-current technology, electrical contacts of palladium and its alloys are used. Large numbers of so-called reed contacts (silver–palladium-, rhodium- or ruthenium-coated contacts) have been used in telephone relays. Palladium can sometimes replace gold in coatings for electronics, electrical connectors and lead frames of semiconductors (Kroschwitz, 1996). The plating solutions contain palladium(II) diamminedinitrite [$Pd(NH_3)_2(NO_2)_2$], the tetra-ammine complex or palladium(II) chloride (Smith et al., 1978; Renner, 1992; Kroschwitz, 1996).

3.2.3.2 Dental materials and other medical materials

Palladium has major importance in dentistry in both cast and direct fillings. Palladium is a component of some dental amalgams. Dental casting gold alloys containing PGMs have been considered the standard material for all types of cast restorations. Palladium alloys (gold–silver–copper–PGM) can be matched to any dental application (inlays, full-cast crowns, long-span bridges, ceramic metal systems and removable partial dentures) by small variations of the alloy composition (Stümke, 1992). For example, there are more than 90 existing palladium alloys, with more than 50% in Germany for fixed restorations with ceramic veneer (Zinke, 1992; Daunderer, 1993). However, in Germany, dentists have recently been advised not to use palladium–copper alloys unless the alloys have been previously tested for corrosion resistance and biocompatibility (Zinke, 1992; BGA, 1993).

Recently, [103]Pd has been used for cancer (e.g., prostrate) brachytherapy, a form of cancer radiation therapy in which radioactive sources are implanted directly into a malignant tumour (Sharkey et al., 1998; Finger et al., 1999).

3.2.3.3 Automobile exhaust catalysts

For more than 20 years, automobile exhaust catalysts have been used to reduce levels of nitrogen oxides, carbon monoxide and hydrocarbons in automobile exhausts. In the last few years, catalysts employing precious metal combinations of platinum or palladium and rhodium in a ratio of 5 to 1 (1.4–1.8 g PGM/litre catalyst volume) have been developed successfully (Abthoff et al., 1994; Degussa, 1995; Kroschwitz, 1996). Exhaust gas purification by equipping of passenger car diesel engines with palladium oxidation catalysts has been achieved only since about 1989 (Fabri et al., 1990), but more recent information shows that palladium is not used on diesel vehicles, which account for around 23% of the European market (Cowley, 1997). Concentrations of the precious metals vary and depend upon the specifications of the manufacturer (IPCS, 1991). Much of this information is proprietary.

Worldwide demand for palladium in automobile catalysts rose from 23.5 tonnes in 1993 tonnes to 76.4 tonnes in 1996 (see Table 8). Around 60% of European gasoline cars sold in 1997 were equipped with palladium-based catalysts. North American car makers continued to use platinum-rich underbody catalysts, but there was increasing use of palladium starter catalysts to meet the hydrocarbon limits imposed by low-emission vehicle legislation. Many Japanese cars are equipped with palladium systems, whereas platinum-rich technology remains dominant elsewhere in Asia (Cowley, 1997).

3.2.3.4 Catalysts in chemical processes

Palladium has a strong catalytic activity for hydrogenation, dehydrogenation, oxidation and hydrogenolysis reactions. Industrial palladium catalysts are in the form of finely divided powder, wire or gauze or supported on substrates such as activated carbon, gamma-aluminium oxide or aluminium silicates. Often, two or more PGMs are combined (Table 9). In the petroleum industry, PGM catalysts are used to produce gasolines with high antiknock properties. Palladium(II) chloride and tetrachloropalladic(II) acid are important homogeneous catalysts used in the large-scale oxidation of ethylene to acetaldehyde in the Wacker process. Palladium catalysts are also used for the acetoxylation of ethylene to vinyl acetate (Fishbein, 1976) and in the manufacture of sulfuric acid and methanol (Smith et al., 1978; Kroschwitz, 1996).

Table 9. Examples of the catalytic activity of palladium[a]

Principal metal	Additional metal	Reaction
Pt, Pd, Ir	Au	oxidative dehydrogenation of alkanes, *n*-butene to butadiene, methanol to formaldehyde, dehydrogenation of alkylcyclohexanes, isomerization and dehydrogenation of alkylcyclohexanes or alkylcyclopentanes, hydrogenative cleavage of alkanes, dealkylation of alkylaromatics
Pd (powder form)	Sn, Zn, Pb	selective hydrogenation of alkynes to alkanes
Pd	Ni, Rh, Ag	alkane dehydrogenation and dehydrocyclization

[a] Adapted from Renner (1992).

3.2.3.5 Fine jewellery and (optical) instruments

The use of palladium for jewellery, coinage and investment has recently begun on a small scale ("white gold"). Palladium's major role in jewellery fabrication is as a subsidiary alloying component of the platinum alloys used in Japan (Coombes, 1990). Alloys are also used for bearings, springs and balance wheels in watches and for mirrors in astronomical instruments. In jewellery, palladium hardened with 4–5% ruthenium provides a light, white, strong, tarnish-free alloy for watch cases, brooches and settings for gems (Budavari et al., 1996; Kroschwitz, 1996).

3.2.4 Uses of important palladium compounds

Uses of some important palladium compounds are described briefly below:

▸ *Ammine complexes of palladium*: The compounds and reactions are important in the industrial separation of palladium, i.e., chloropalladosamine is a precurser of metallic palladium sponge. It is also used in electroless plating and bright palladium plating. Ammonium hexachloropalladate(IV) is important in separation technology.

▸ *Palladium(II) acetate*: Palladium(II) acetate is of some importance in preparative chemistry. It is used as a catalyst (Budavari et al., 1996).

▸ *Palladium(II) chloride*: Palladium(II) chloride is used in plating baths. Pellets or monoliths of oxidation catalysts are either immersed in an aqueous solution of palladium(II) chloride (impregnation technique) or sprayed with a solution of this chemical (NAS, 1977).

Other uses for palladium(II) chloride include photography, toning solutions, electroplating parts of clocks and watches, detecting carbon monoxide leaks in buried gas pipes, manufacture of indelible ink and preparation of metal for use as a catalyst (Budavari et al., 1996; Olden, 1997). Different purity grades of palladium(II) chloride ranging from 99% to 99.999% are available for chemical or medical use (Aldrich, 1996).

▸ *Palladium(II) nitrate*: Palladium(II) nitrate is used as a catalyst in organic syntheses and in the separation of chlorine and iodine (NAS, 1977; Budavari et al., 1996).

▸ *Palladium(II) oxide*: Palladium(II) oxide is used as a hydrogen-ation catalyst in the synthesis of organic compounds.

▸ *Hydrogen tetrachloropalladate(II)*: The solution of hydrogen tetrachloropalladate(II) is an industrially important palladium preparation. It is the starting material for many other palladium compounds, particularly catalysts (Renner, 1992).

▸ *Tetraammine palladium hydrogen carbonate*: Tetraammine palladium hydrogen carbonate is used as an intermediate in the production of automobile catalysts (Lovell, Johnson Matthey plc, personal communication, February 2000).

3.3 Emissions during production and use

Since palladium is a valuable metal, great care is taken to avoid significant loss during mining and refining processes and during use and disposal of palladium-containing objects.

There are no data available concerning losses of palladium to the atmosphere and potentially to aquatic sinks from the use of catalysts in the chemical and petroleum industry. Used palladium catalysts can be recovered with a loss of about 5–6% (Fishbein, 1976).

3.3.1 Emissions into air

3.3.1.1 Production and fabrication losses

There are three major categories of industrial point sources for possible emission of palladium compounds: mining (where no infor-mation is available), refining and processing.

Older data are available on the annual amount of palladium lost for the smelter stacks in Sudbury, Ontario (Canada). Palladium losses of 69 kg were reported for the year 1971 (Smith et al., 1978). The data show that palladium can be lost in particulate and gaseous emissions during smelting of copper, nickel and other base metal ores containing PGMs. Other data on emissions of palladium during production are not available.

Also, during the use of stationary palladium-containing catalysts, palladium may escape into the environment, but there is no measurement available to support this assumption.

3.3.1.2   *Losses from automotive exhaust emission control catalysts*

Experimental data show that automobile catalysts are likely to emit palladium into the environment. These emissions may be due to mechanical and thermal impact. Information on the palladium emission rate of cars equipped with modern monolithic palladium/rhodium three-way catalysts is still scarce.

Moldovan et al. (1999) determined PGM concentrations in the raw car exhaust fumes released by two different types of fresh gasoline catalytic converters (platinum/palladium/rhodium and palladium/rhodium), a diesel catalyst (labelled as platinum only) and an 18 000-km aged platinum/palladium/rhodium gasoline catalyst. Palladium was the main noble metal in the three gasoline catalysts. Samples were collected following 91 441 urban and extra-urban driving cycles for light-duty vehicle testing. As shown in Table 10, the particulate palladium released from the fresh catalysts is in the range of 3.7–108 ng/km.

Table 10. Particulate and soluble palladium emissions in exhaust fumes from new catalysts

	Emissions (ng/km)		
	Pt/Pd/Rh catalyst	Pd/Rh catalyst	Pt diesel catalyst
Particulate	93–108	18.5–23.2	3.7–6.0
Soluble	12.3–14.1	4–7.8	not detected
(% of total)	(10.2–13.2)	(15.7–29.7)	–

Most of the palladium released was in particulate form. The soluble fraction represents 10–30% of the total released. In a preliminary study performed on the aged platinum/palladium/rhodium catalyst, the palladium emission was lower at a constant speed of 80 km/h (1.2–1.9 ng/km) than when the driving cycle was applied (2–24 ng/km). Hence, new catalysts seemed to release more particulate palladium than older catalysts.

According to experimental data with a laboratory-prepared pellet-type catalyst containing both [195]Pt and [103]Pd, emissions are approximately 1.1 µg palladium/km, with an exhaust gas concentration of approximately 0.48 µg palladium/m^3. About 10% of the active material lost is water soluble, possibly as halides. Eighty per cent of the particles collected had diameters greater than 125 µm (Hill & Mayer, 1977). It should be noted that this type of automobile catalyst is no longer used in new cars and that with the new monolith-type catalysts, the emission may be lower by a factor of 100 (see section 3.3.1.3). The results relate to the first 250 km of catalyst life, and lower loss rates would be expected with increasing age of the catalyst.

In a more recent study, samples of PGMs were taken from a standard-type three-way platinum/palladium/rhodium catalyst-equipped gasoline engine, running on a computer-controlled dynamometer (Lüdke et al., 1996). The measured concentration for palladium was 0.3 ng/m^3 (platinum, 120 ng/m^3).

Characterization of heavy-duty diesel vehicle emissions was done by Lowenthal et al. (1994). Surprisingly, there was no significant difference in the palladium emissions for trucks and buses with and without particulate traps. Palladium emission rates for $PM_{2.5}$ (particulate matter with aerodynamic diameter <2.5 µm) were between 6 and 24 µg/km. These values are questionable, because the uncertainty of the applied X-ray fluorescence analysis was very high, and it is not evident from the study whether the filters contained palladium or whether the emissions were related to the examined different fuel types.

With the exception of this study, nothing is known about the fraction of free palladium particles <3 µm, which would be in the respirable size range.

3.3.1.3 Experimental results with platinum-type automobile catalysts to estimate palladium emissions

In a first approximation, experimental results with platinum-type automobile catalysts may be considered to estimate palladium emissions. In the case of platinum catalysts, platinum is emitted into the environment in highly disperse, mainly metallic form in conjunction with the gamma-aluminium oxide washcoat that serves as a carrier for the metal (Helmers & Kümmerer, 1997).

Experimental tests with a 1.6-litre, 51-kW engine, operated under simulated stationary speed conditions, gave total observed platinum concentrations ranging from 80 to 110 ng/m³ (Inacker & Malessa, 1997). Further experiments with different engines (1.8-litre, 66-kW; 1.4-litre, 44-kW) run at comparable speeds showed concentrations ranging from 8 to 87 ng platinum/m³. Of the particles collected, 33–57% had aerodynamic diameters <10 μm and 10–36% had diameters <3.14 μm (Artelt, 1997). Taking into account the measured gasoline consumption for the simulated speed conditions, the determined concentrations in the exhaust gas correspond to emission rates of approximately 60–87 ng/km. No further details of the calculation method were given in Artelt (1997).

If one assumes that palladium particles behave similarly to platinum particles, the conclusion may be drawn that there are palladium particles emitted from the catalytic converter in the respirable range, which could reach the lungs.

3.3.2 Emissions into water

No data are available for loss of palladium(II) compounds from refineries and electroplating processing into wastewater streams.

3.3.3 Emissions into soil

Data for soil are not available. Release of sewage sludges into the environment may cause contamination of soil (see section 5.1.4).

4. ENVIRONMENTAL TRANSPORT, DISTRIBUTION AND TRANSFORMATION

4.1 Transport and distribution between media

Rain may wash the palladium particles concentrated in roadside dust into local water systems or the ocean (Hodge & Stallard, 1986).

The chemical nature and thermodynamics of palladium minerals and aqueous species suggest that palladium is mobile as hydroxide, chloride and bisulfide complexes, depending on pH, oxygen fugacity, ligand concentration and temperature. Hydroxide complexes appear to predominate over chloride complexes at near-neutral to basic pH, even at high chlorinities (5 mol Cl^-/litre). The data suggest that the predominant inorganic form of palladium in fresh waters may be the neutral hydroxide species. Bisulfide complexing ($[Pd(HS)_4]^{2-}$) and mixed bisulfide–hydroxide complexes may be important under similar pH conditions. In seawater (near pH 8), it is currently not clear whether the chloride or the hydroxide complex will predominate (Wood, 1991). The dominant chloride species is $PdCl_4^{2-}$. Complexes of palladium with carbonate, bicarbonate, phosphate, sulfate and fluoride ions are predicted to be weak (Mountain & Wood, 1987, 1988).

The solution chemistry of palladium in seawater has been examined by UV absorbance spectroscopy (Kump & Byrne, 1989). The results indicate that chloride complexation strongly suppresses the hydrolysis of Pd^{2+} in seawater. At 25 °C and 36‰ salinity, hydroxy complexes account for approximately 18% of the total palladium, and the remainder is present principally as $PdCl_4^{2-}$.

The complexation of palladium(II) by amino acids (e.g., glycine: G^-) appears to play a significant role in the metal's solution chemistry (Li & Byrne, 1990). At 25 °C, 35.3‰ salinity and very low concentrations of free unprotonated glycine (3 nmol/litre), the formation constant for the mixed species $PdCl_2(NH_2CH_2COO)^-$ is given as $\beta = 10^{19.6}$. These results indicate that $PdCl_2G^-$ complexes are more stable than all known inorganic complexes of palladium(II).

Palladium powder (particle size not determined) was dissolved at room temperature in oxygen-saturated 0.2 mol/litre solutions of biogenic compounds (α-amino acids and peptides) in water at pH 5–7, and the solutions were stirred for 3 weeks. The resulting mass concentrations of dissolved palladium ranged from 21 to 42 µg/ml. Under the same conditions, the value for pure water is only 0.42 µg/ml (Freiesleben et al., 1993). Organic matter in polluted stream sediments has long been studied for its metal-complexing capacities. Dissanayake et al. (1984) outlined the probable importance of humic and fulvic acids in the transport of palladium.

Fuchs & Rose (1974) studied the geochemical behaviour of palladium in the weathering cycle of the Stillwater complex, Montana (USA). Their study of palladium distribution among the exchangeable, organic, iron oxide, clay, silt, sand and magnetic fractions of four soils showed that considerable palladium occurs in the clay fraction and the organic component. On the basis of the high palladium concentration in a limber pine (*Pinus flexilis*), they concluded that palladium is mobile in the organic cycle (see section 5.1.7.1).

4.2 Transformation and removal

4.2.1 Abiotic

Most of the palladium in the biosphere is in the form of metal or metal oxides, which are almost insoluble in water, are resistant to most reactions in the biosphere (e.g., abiotic degradation, UV radiation, oxidation by hydroxyl radicals) and do not volatilize into air (Renner & Schmuckler, 1991).

4.2.2 Biotic

As a metal, palladium is not biodegradable, but it was assumed to undergo methylation reactions (Fishbein, 1976; Johnson et al., 1976; Morgan & Stumm, 1991). There are no data to support this. If this occurs, palladium could be concentrated along the food-chain.

In vitro salts of palladium were observed to demethylate methylcobalamin, a biologically active form of vitamin B_{12}, at pH 2.0, but the rates are much slower than are observed with Pt^{4+} salts (Taylor, 1976).

The three palladium compounds shown to be reactive were as follows: $K_2PdCl_6 > PdSO_4 > K_2PdCl_4$.

However, no evidence for a stable methyl palladium derivative was found. Furthermore, there is no evidence that palladium can be methylated unless it is in the 4+ or 2+ valence.

4.3 Bioaccumulation

4.3.1 Aquatic organisms

Bioconcentration factors ranging from 5000 to 22 000 have been calculated for marine macrophytes (Yang, 1989). Interpretation of these data is limited, as the bioconcentration factors were not determined with reference to measured exposure concentrations. They were derived by comparing concentrations measured in macrophyte fronds (see Table 13 in chapter 5) with a nominal palladium concentration found in northeast Pacific Ocean water of 0.02 ng/litre. The authors did not state the exposure period or whether the bioconcentration factors were expressed on a wet or dry weight basis.

Palladium ions were taken up by the aquatic plant water hyacinth (*Eichhornia crassipes*) and accumulated in the roots (Farago & Parsons, 1994). The plants were grown in a culture medium, and soluble palladium complexes were added at a concentration of 2.5 mg/litre. After 2 weeks, the plants were harvested, dried, ashed and analysed for palladium by flame AAS and flameless electrothermal atomization AAS. Concentrations in dry plant material of 20 mg/g (roots) and 1 mg/g (tops) were measured.

Palladium concentrations of 133 and 785 µg/kg dry weight, respectively, were found in two samples of the leaves of the water hyacinth (*E. crassipes*) growing in water contaminated with palladium (Abbasi, 1987).

4.3.2 Terrestrial organisms

There are no data available on the bioaccumulation of palladium in terrestrial organisms.

4.4 Ultimate fate following use

Although some catalytic converters or electronic components as well as dental alloys are recycled, it can be assumed that some products containing palladium are disposed of in landfills or waste dumps (see section 3.2.1.2). No prognosis of potential palladium releases is possible, because nothing is known about the mobilization processes for palladium.

5. ENVIRONMENTAL LEVELS AND HUMAN EXPOSURE

5.1 Environmental levels

5.1.1 Air

There are very few data on palladium concentrations in the atmosphere or on its chemical form.

In 1974 (before the introduction of automotive catalytic converters), the palladium concentrations in an area in California (USA) with high traffic density but unpolluted with palladium were below the detection limit (0.06 pg palladium/m^3) (Johnson et al., 1976).

In 1992–1993, temporal and spatial variations and chemical constituents of PM_{10} aerosol (particulate matter with aerodynamic diameter <10 µm) in Imperial County, California (USA), were examined by Lu et al. (1994). Observed PM_{10} concentrations averaged 52.2 µg/m^3, with the average palladium concentration not exceeding 1 pg/m^3 ($n = 118$).

$PM_{2.5}$ aerosols were sampled at Caesarea (Israel) from May to June 1993, a time of year when air mass transport from Europe is frequent (Gertler, 1994). Observed $PM_{2.5}$ concentrations in 10 samples averaged 25.5 µg/m^3, with palladium accounting for 3.3 pg/m^3.

An ambient air study was conducted in the city of Chernivtsi (Ukraine) from 26 October to 2 November 1990 to measure total particulate matter and the elemental composition, including palladium (Scheff et al., 1997). The results were compared with studies in Chicago, Illinois (USA). The average total particulate matter concentration (144 µg/m^3) and palladium concentration (56.6 ng/m^3) measured in Chernivtsi exceeded the values reported in Chicago (PM_{10}: 30.3 µg/m^3; palladium: 12.7 ng/m^3). The reported palladium concentrations are in the same order of magnitude as typical values for transitional elements like manganese, zinc or copper and were allocated to "other anthropogenic activity," because soil and automobile contributions were assumed to be negligible.

46

Platinum concentrations, measured or calculated, may be considered to be rough approximations of palladium concentrations. Ambient air levels of platinum were predicted by applying dispersion models including recent emission factors derived from engine test bench experiments with three-way monolith-type converters (Rosner et al., 1998). The calculated concentrations ranged from 4 pg/m^3 (street canyon, typical conditions) up to 112 pg/m^3 (express motorway, severe conditions). These values agree with the few measurements of platinum air concentrations. For example, ambient air concentrations of platinum between 8 and 106 pg/m^3 have been reported from Munich (Germany) in areas with high traffic density where platinum catalysts are used.

Based on these sparse data, it is impossible to make a generalization about the concentrations of palladium in air.

5.1.2 Dust

A summary and literature overview of actual palladium concentrations and platinum/palladium ratios in dust are given in Table 11.

Roadside dust samples collected from broad-leaved plants in San Diego, California (USA) (1985) contained 15–280 µg palladium/kg, 10% of it as dissolved forms (Hodge & Stallard, 1986). The highest values were from highways with heavy traffic, and the lowest (15 µg/kg) were from plants in gardens on streets with light traffic. The average platinum/palladium ratio was 2.5.

Palladium concentrations of 1–146 µg/kg were found in urban road dusts in Germany in 1995. The values were strongly correlated to the density of traffic and exceeded natural concentrations, which are below 0.8 µg/kg (Schäfer et al., 1996).

The palladium concentration in US NBS primary standard coal fly ash (SRM 1633) was found to average 1.8 ± 0.3 µg/kg (Shah & Wai, 1985).

5.1.3 Soil

Actual values for palladium concentrations in soil are summarized in Table 11. In an area in California (USA) with high traffic density and unpolluted with palladium, the palladium concentration was below

Table 11. Palladium in environmental matrices

Matrix	Date of sampling	Pd concentration (µg/kg)	Pt/Pd ratio	Remarks	Reference
Dust					
Traffic tunnel dust (Munich, Germany)	1994	20	9.5		Helmers et al. (1998)
Traffic tunnel dust (Japan)	1987	297		mainly soot (80% carbon)	Helmers et al. (1998)
Dust near German highway	June 1995	1–146		concentrations depend on the density of traffic	Schäfer et al. (1996)
Urban roadside dust (Frankfurt, Germany)	August–October 1994	5 (mean)	10		Zereini et al. (1997)
Roadside dust (heavy traffic, San Diego, California, USA)	February/June 1985	115 (mean)	2.6		Hodge & Stallard (1986)
Roadside dust (light traffic, San Diego, California, USA)	July 1985	19.5 (mean)	2.5		Hodge & Stallard (1986)
Soil					
Soil near Californian highway	1975	<0.7		background value	Johnson et al. (1976)
Soil near German highway	October 1990 – March 1991	1–27 (mean: 2)	5	0- to 4-cm layer; distance to road: 1 m	Zereini et al. (1997)
Soil near German highway	August–October 1994	1–47 (mean: 6)	11	0- to 4-cm layer; distance to road: 1 m	Zereini et al. (1997)

Table 11 (contd).

Matrix	Date of sampling	Pd concentration (µg/kg)	Pt/Pd ratio	Remarks	Reference
Soil near German highway	August–October 1994	0.7	11	0- to 4-cm layer; distance to road: 10 m	Zereini et al. (1997)
Soil near German highway	June 1995	1–10	6	0- to 2-cm layer	Schäfer et al. (1996)
Sludge					
Municipal sewage sludges (Australia)	1995	18–153		samples from heavily polluted areas	Lottermoser (1995)
Muncipal sewage sludges (Germany)	1995	260 4700		median value noble metal-bearing discharges	Lottermoser (1995)
Sediment					
<2 µm fraction sediments (Germany)	1984	≤4–4000		samples from a highly polluted cut-off channel	Dissanayake et al. (1984)
Sediment near German highways	1996	0.7–19.3		0- to 2-cm layer; distance to road: 0–1 m	Cubelic et al. (1997)
Sediments from the eastern Pacific	1988	1.3–9.4 (mean: 3.2)		pelagic sediments	Goldberg et al. (1988)
Sediments from the eastern Pacific	1988	0.1–13.7 (mean: 2.9)		anoxic sediments	Goldberg et al. (1988)
For comparison: Geological background					
Continental crust		0.4	1		Wedepohl (1995)

the detection limit (0.7 µg/kg) of the AAS technique used (Johnson et al., 1976).

Measurements taken in 1990 and 1991 along highway (Autobahn) A66 Frankfurt–Wiesbaden in Germany gave a mean value of 2 µg palladium/kg (Zereini et al., 1993). The use of palladium for catalytic converters in automobiles has greatly increased since 1992. More recent measurements in 1994 of the top layer of soil along highway A67 Frankfurt–Mannheim gave a mean value of 6 µg palladium/kg (Zereini et al., 1997). Only the soil layers down to 20 cm depth contained measurable contents of palladium. Levels of palladium decreased with increasing distance from the edge of the roadway.

To date, only one surface soil analysis has been made in connection with palladium production. In 1974, the palladium concentration in the area around a mine in Sudbury, Ontario (Canada), where PGMs are exploited, was determined to be 2.0–4.5 µg/kg (Johnson et al., 1976).

5.1.4 Sludges

Municipal sewage sludges in southeastern Australia contain palladium concentrations of 0.018–0.153 mg/kg dried material from industrialized and heavily polluted areas (Lottermoser, 1995) (Table 11). The palladium content in selected German sewage sludges was about 0.26 mg/kg dry weight. The high palladium value (4.7 mg/kg) in a sample of sewage sludge from Pforzheim (Germany) was due to noble metal-bearing discharges from the local jewellery industry (Lottermoser, 1995). Urban sludges receive emissions not only from traffic, but also from local industrial activities.

5.1.5 Sediments

Dissanayake et al. (1984) determined palladium concentrations in the <2 µm fraction sediments of a highly polluted cut-off channel of the river Rhine near Mainz (Germany). The palladium concentrations in 12 samples collected at different sites varied over a wide range. Eight samples did not have any detectable quantities of palladium (i.e., <4 µg/kg), while four samples contained 30 µg/kg, 50 µg/kg, 410 µg/kg and the extremely high concentration of 4.0 mg/kg dry weight (Table 11). The palladium and platinum content correlated very well with each other and also with organic carbon.

In pelagic sediment cores from the eastern Pacific Ocean taken to a depth of 6–22 cm, palladium concentrations ranged between 1.3 and 9.4 µg/kg (mean value 3.2 µg/kg dry weight). For anoxic sediments, values between 0.1 and 13.7 µg/kg (mean value 2.9 µg/kg) were reported (Goldberg et al., 1988).

5.1.6   Water

5.1.6.1   Fresh water

Water samples from streams and ponds were collected in and around the mining and ore processing facilities located in Sudbury, Ontario (Canada) (Johnson et al., 1976). All samples were below the detection limit of 15 ng palladium/litre.

Shah & Wai (1985) found concentrations of 0.4 ± 0.1 ng palladium/litre in natural waters collected from the Snake River in Idaho (USA). Palladium concentrations of 1.0 ± 0.1 ng/litre (river Schwarzbach) and 0.4 ± 0.1 ng/litre (river Rhine) were determined for natural waters in Germany (Eller et al., 1989). Palladium concentrations at selected sites in Ferguson Lake and Rottenstone Lake (Canada) were 1–2 ng/litre (Hall & Pelchat, 1993). Some samples were below the detection limit of 0.5 ng/litre. Stetzenbach et al. (1994) found measurable amounts of palladium (≤3 ng/litre) in water samples from four springs in Nevada (USA). A concentration of 22 ng palladium(II)/litre was found in a spring water sample collected in 1990 in Osaka (Japan) (Chikuma et al., 1991).

A palladium concentration of 0.1 µg/litre was determined in spring water in the People's Republic of China (Zhou & Liu, 1997).

Palladium analysis of rain collected in simple open collectors in Stuttgart (Germany) in July 1996 gave concentrations below the detection limit of 5 ng/litre (Helmers et al., 1998).

5.1.6.2   Seawater

Palladium concentrations of 22 pg/litre (depth 0–10 m) and 60 pg/litre (depth 3000 m) were measured by Goldberg (1987) in samples of filtered Pacific Ocean waters. Palladium probably exists as a complex form of its divalent state in seawater. Goldberg et al. (1988) found average concentrations of about 40 pg palladium/litre (mainly as $PdCl^+$ ions in the 2+ state) in surface seawater. Palladium profiles in marine water and sediment were also reported by Lee (1983). The palladium concentration in the water column of the northeast Pacific Ocean increased from 19 pg/kg at the surface to 70 pg/kg in deep waters. A strong co-variance between palladium and nickel implies similar biogeochemical pathways.

5.1.7   Biota

5.1.7.1   Plants

1)   Terrestrial plants

Palladium was analysed in a standard grass culture near a German highway with high traffic intensity (100 000 cars/day). In all samples (20 and 50 cm distance to road, sampled in 1992–1995), palladium concentrations were found to be lower than 0.5 µg/kg dry weight (Helmers et al., 1998). Soil was not measured at the same time, but palladium concentrations in soil are normally higher than 1 µg/kg (see Table 11).

Kothny (1979) found palladium in the ash of a number of plant species from the Orinda formation in California and showed a seasonal variation in the leaves of California black walnut (*Juglans hindsii*). Other species analysed showing uptake of palladium with an ash/soil factor above 2 included California white oak (*Quercus lobata*), red pine (*Pinus resinosa*), manzanita (*Arctostaphylos nummularia*) and canyon live oak (*Quercus chrysolepis*) (Table 12). Kothny (1979) suggested that palladium(II) forms part of a metalloenzyme or that it can replace manganese(II), since it has a similar ionic radius.

Fuchs & Rose (1974) analysed ashed samples of twigs from four limber pines (*Pinus flexilis*) in the Stillwater mining area, Montana

Table 12. Palladium in plant ash[a,b]

Species and residual soil	Mean ± SD (µg/kg)	Ratio ash/soil	Ash %[c]
Soil 1	40 ± 20 (9)[d]	–	–
Picea pungens[e]	30 ± 0 (2)	0.8	4
Quercus pungens[e]	110 ± 30 (2)	2.7	6
Prunus domestica[e]	50 ± 10 (3)	1.3	10
Pinus resinosa[f]	110 ± 10 (2)	2.7	4
Pinus radiata[f]	15 ± 15 (2)	0.4	3
Tragopogon porrifolias[g]	80 ± 30 (4)	2	7
Lactuca virosa[h]	50 ± 20 (2)	1.3	7
Populus fremontia[h]	40 ± 5 (2)	1	10
Arctostaphylos nummularia[e]	110 ± 30 (2)	2.7	3
Eucalyptus sp.[h]	35 ± 15 (3)	0.9	10
Juglans hindsii[i]	52 ± 28 (30)	1.3	12
Soil 2	140 ± 20 (2)	–	–
Quercus chrysolepsis[e]	400 ±100 (2)	2.8	2.4[j]

[a] Analytical results on ash basis; SD = standard deviation.
[b] Compiled from Kothny (1979).
[c] From air-dried samples at 50% room humidity and 25 °C.
[d] Number of determinations in parentheses.
[e] Leaves and tips.
[f] Needles and tips.
[g] Whole plant.
[h] Leaves.
[i] Leaves and stalks.
[j] Wet basis.

(USA) (see section 4.1). Three samples contained between 2 (till area) and 11 µg palladium/kg ash, while an anomalous 285 µg palladium/kg ash was found in the sample from a mineral deposit area. The data indicated that palladium may be transported into limber pine in significant amounts (Fuchs & Rose, 1974).

The concentration and distribution of palladium were studied in two arctic shrub species, Labrador tea (*Ledum palustre*) and dwarf birch (*Betula nana*), growing in the vicinity of a PGM mineral deposit area at Ferguson and Townsend lakes in Canada (Rencz & Hall, 1992).

Depending on the distance from the mineral deposit, palladium concentrations in twigs from dwarf birch ranged from 8.5 (background) to 4014 µg/kg ash, and concentrations in leaf tissue of Labrador tea ranged from 37.8 (background) to 2523 µg/kg ash.

Biomonitoring values from different areas of Italy using epiphytic lichens as bioaccumulators were below the detection limit of the inductively coupled plasma atomic emission spectrometry (ICP-AES) technique of 2.5 µg/kg (Guidetti & Stefanetti, 1996; Zocchi et al., 1996).

Palladium concentrations below or at the limit of detection (0.25 µg/kg dry weight) were found in the US NBS standard reference sample of orchard leaves (SRM 1571) (Eller et al., 1989).

The uptake of traffic-related PGMs by plants has been determined by Schäfer et al. (1998). Palladium concentrations in spinach, cress and phacelia grown on palladium-contaminated soil (5.7 and 6.5 µg palladium/kg) collected from areas adjacent to a German highway were close to the detection limit of the AAS method (~0.7 µg palladium/kg dry material), whereas concentrations in stinging nettle (*Urtica dioica*) reached up to 1.9 µg palladium/kg.

In 1995, palladium concentrations in grass from sampling locations near German highways were between <0.5 and 0.6 µg palladium/ kg dry weight (Hees et al., 1998).

2) Aquatic plants

Yang (1989) analysed annual fronds from 22 species of marine macrophytes collected from California (USA) in 1986 using GF-AAS. Mean palladium concentrations for the different types of algae are listed in Table 13.

5.1.7.2   *Animals*

Concentrations of palladium were quantified in body segments of different size classes of the spot prawn (*Pandalus platyceros*) collected from various locations in British Columbia (Canada). The concentration of absorbed palladium was highest in the abdomen carapace

Table 13. Average palladium concentrations in seaweeds[a]

Algal type	Concentration (µg/kg)
Brown	0.11
Non-calcareous red	0.44
Calcareous red	0.17
Green	0.27

[a] Table adapted from Yang (1989).

(6 mg/kg dry weight) and the head (5.64 mg/kg dry weight). The lowest concentration (0.02 mg/kg dry weight) was found in the hatched larvae. The data of this field sampling study suggested an accumulation of palladium during the growth and development of marine invertebrates (Whyte & Boutillier, 1991).

In muscles of feral pigeon (*Columba livia*), a concentration of 0.511 mg palladium/kg dry weight was measured (Abbasi, 1987).

5.2 Exposure of the general population

5.2.1 Levels found in the general population

Exposure of the general population is through palladium in air, food and water and through release of palladium from dental restorations containing palladium.

At present, the low detection sensitivity for analysis in biological matrices is still a problem. If current tissue data vary by a factor of 2 or more from older data, this can reflect different trace levels as well as the analytical reliability of the detection method.

In earlier reports, palladium concentrations in serum, blood, faeces, hair and urine of persons without occupational exposure were below the detection limits (Gofman et al., 1964; Johnson et al., 1975b, 1976). A pool of 282 blood samples indicated concentrations of less than 0.01 µg/100 ml blood for palladium (Johnson et al., 1975b). More recent background levels for urine do not show a general concentration trend in the overall population over the last 25 years. Some levels of palladium in blood, hair, urine and faeces in non-occupationally exposed persons are shown in Table 14.

Measured concentrations of palladium in various tissues were reported in a study completed in 1974 (before the introduction of catalytic converters). Concentrations in autopsy "wet" tissue samples (liver, kidney, spleen, lung, muscle, fat) from five men and five women 12–79 years of age who died from a variety of causes in Los Angeles, California (USA), were below the limit of detection for different tissue materials by the applied AAS technique — i.e., less than 0.6–6.7 µg/kg (Johnson et al., 1976).

The investigations of Begerow et al. (1998a, 1999b) showed that direct exposure to traffic has no verifiable influence on the background burden of the population.

It is probable that the baseline concentrations of palladium in the body fluids of unexposed people are <0.1 µg/litre blood and ≤0.3 µg/litre urine (Table 14).

5.2.2   Food

Little information was found in the literature, and only one total diet study was available. The actual concentrations of palladium in food from various countries vary widely depending on the food product and the growing conditions (soil).

The following mean palladium concentrations per kg wet weight were measured in different meat samples: chicken, 0.6 µg/kg; beef, 0.7 µg/kg; and pork, 0.5 µg/kg. The highest concentration of 2 µg palladium/kg was in ham (Koch & Roesmer, 1962).

A concentration of 0.5 µg palladium/kg was found in a US NBS standard reference sample of bovine liver (SRM 1577) (Eller et al., 1989).

Nineteen honey samples collected from several polluted areas in Florida and New York State (USA) were analysed before the introduction of catalytic converters in the USA (Tong et al., 1975). The background concentration range of palladium was <1–15 µg/kg fresh

Table 14. Palladium concentrations in different tissue samples of non-occupationally exposed men and women

Sample	Year	Place	Number in group; details	Concentration[a] (range)	Reference
Blood	1975	southern California	282 (pooled)	<0.1 µg/litre[b]	Johnson et al. (1975b)
Hair	1975	southern California	282	<0.02 µg/g[b]	Johnson et al. (1975b)
Faeces	1975	southern California	282	<0.001 µg/g[b]	Johnson et al. (1975b)
Urine	1975	southern California	282	<0.3 µg/litre[b]	Johnson et al. (1975b)
Urine	1989	Italy	136; 24-h samples	<0.15 µg/litre	Minoia et al. (1990)
Urine	1994	Germany	human biomonitoring reference value	<0.1 µg/litre	Schaller et al. (1994)
Urine	1997	Germany	14; 24-h samples (non-smokers)	0.03–0.2 (mean 0.1) µg/litre	Schramel et al. (1997)
Urine	1995	Germany	9; 24-h samples	0.02–0.08 (mean 0.04) µg/litre	Begerow et al. (1997a)
Whole blood	1996	Germany	7	0.03–0.08 (mean 0.05) µg/litre	Begerow et al. (1997b)
Urine	1996	Germany	21; 24-h samples	0.03–0.22 (mean 0.14) µg/litre	Begerow et al. (1997c)
Urine	1998	Germany	17, morning urine	0.013–0.048 (mean 0.031) µg/litre	Begerow et al. (1999b)
Urine	1998	Germany	262 (total number) exposure to low traffic: 92 medium traffic: 83 high traffic: 86, morning urine samples[c]	0.006–0.091 (mean 0.033) µg/litre; mean values (µg/litre): low traffic, 0.033; medium traffic, 0.035; high traffic, 0.032	Begerow et al. (1998)

[a] < = Lower than the detection limit.
[b] Baseline levels of palladium in populations of southern California (1975) prior to the introduction of automobile catalytic converters. Measurements based on AAS.
[c] Children without dental appliances (age 7–10) from areas with low–high traffic density.

weight. Palladium levels in two samples of honey produced by bees near a motorway were 7 and 9 µg/kg fresh weight, respectively.

Palladium concentrations in various foodstuffs in the People's Republic of China (1996–1997) were given as follows (in µg/kg): peanut oil, 0.9; corn, 1; orange juice, shaddock, rice flour, 2; beef, chicken stomach, pork liver, raw sugar, 3; red wine, 6; beltfish, celery, 9; mussels, 46; and tea, 80 (Zhou & Liu, 1997).

The United Kingdom's Ministry of Agriculture, Fisheries and Food carried out a total diet study in 1994 to identify trends in dietary intakes by the general population of various constituents, including PGMs (MAFF, 1997, 1998; Ysart et al., 1999). In the total diet study, the major items of the national diet were combined into 20 groups of similar foods for analysis. Foods were grouped so that commodities known to be susceptible to contamination (e.g., offal) were separated, as were commodities that are consumed in large quantities (e.g., bread, potatoes and milk). Multi-element analyses for 30 elements (including PGMs) were carried out using ICP-MS. All samples were analysed in duplicate in separate batches. Each batch contained reagent blanks, spiked reagent blanks and three separate reference materials. All test batches met established criteria with regard to instrument drift, recoveries from spiked blanks, agreement between replicates, limits of detection and comparison with measured certified reference material concentrations. The concentrations of the PGMs are the means of the concentrations found in each food group from the 20 towns included in the 1994 total diet study.

Concentrations of the PGMs palladium and platinum included in the 1994 total diet study were low in two food groups (milk and poultry), which contained palladium concentrations below the limit of detection of 0.3 µg/kg (Table 15). Those food groups that contained detectable mean concentrations of PGMs were nuts (3 µg/kg) and fish, offal and bread, which all had mean palladium concentrations of 2 µg/kg. Concentrations in all other food groups were close to the limit of detection. Palladium concentrations were in general higher than the corresponding platinum concentrations.

Dietary intake estimates (i.e., the relative proportion of each food listed in Table 15) are based on data from the United Kingdom National Food Survey. Applying these quantities to the mean

Table 15. Mean concentrations of the PGMs palladium and platinum included in the 1994 total diet study

Food group	Mean concentrations (µg/kg fresh weight)		Food group	Mean concentrations (µg/kg fresh weight)	
	Pd[a]	Pt[a]		Pd[a]	Pt[a]
Bread	2	0.1	Green vegetables	0.6	0.1
Miscellaneous cereals	0.9	0.1	Potatoes	0.5	<0.1
Carcass meat	0.4	0.1	Other vegetables	0.5	0.1
Offal	2	<0.1	Canned vegetables	0.4	<0.1
Meat products	0.6	<0.1	Fresh fruit	0.4	<0.1
Poultry	<0.3	0.1	Fruit products	0.5	<0.1
Fish	2	<0.1	Beverages	0.4	<0.1
Oils and fats	0.4	0.2	Milk	<0.3	<0.1
Eggs	0.4	<0.1	Dairy produce	0.4	0.1
Sugars and preserves	0.5	0.1	Nuts	3	0.1

[a] Limits of detection: palladium, 0.3 µg/kg fresh weight; platinum, 0.1 µg/kg fresh weight.

concentration of each element gives an estimate of population average intake (covering both adults and children). For dietary intakes estimated from results less than the lower limits of detection, it has been assumed that the concentrations of PGMs are at the appropriate limits of detection and are thus upper-bound estimates. Dietary intake estimates of the PGMs for average, mean and upper-range (97.5th percentile) consumers were very low (Table 16).

5.2.3 Drinking-water

Palladium levels in drinking-water, whether distributed through household plumbing or as bottled water, are expected to vary widely according to the natural levels found in the source.

In 1974, the palladium concentrations in the tap water in an urban, palladium-unpolluted area in California (USA), as well as in the tap water in the Sudbury mining (nickel, copper, platinum and palladium) complex (Canada), were reported to be below the detection limit

Table 16. Total average mean and 97.5th percentile dietary intake estimates of palladium and platinum

Element	Total dietary intake (µg/day)		
	Population average[a]	Mean[b]	97.5th percentile[b]
Palladium	1	1	2
Platinum	0.2	0.2	0.3

[a] Population average intakes have been estimated from the mean concentrations of each of these elements in 20 food groups and the average consumption of each food group from the National Food Survey.

[b] Mean and upper range (97.5th percentile) total intakes have been estimated from the mean concentrations of each element in 20 food groups and data on consumption of each food group from the Dietary and Nutritional Survey of British Adults.

(0.024 µg/litre) (Johnson et al., 1976). Tap water in one city in the People's Republic of China contained 0.3 µg palladium/litre (Zhou & Liu, 1997). Detailed information on this single value is not available.

5.2.4   Iatrogenic exposure

5.2.4.1   In vitro studies

For economic reasons, a large number of palladium alternatives to dental casting gold alloys have been introduced on the market. The physical and chemical properties of these alternative alloys have been questioned by some researchers.

Wataha et al. (1991a) showed that palladium release into cell culture medium from a variety of dental casting alloys was not proportional to the atomic percentage of palladium in the alloys. Palladium was non-labile in any tested alloy environment. Palladium was present at levels below 17 µg/litre in the cell culture medium at 72 h for nine different commercial alloys. For one multiphase alloy (Au52, Ni28, Ga13, Pd4, In4; atomic per cent), the palladium concentration in solution after 72 h was 29 µg/litre (0.003 µg palladium/cm^2 per day), but the gallium and nickel concentrations were 8.7 mg/litre (0.97 µg gallium/cm^2 per day) and 14.4 mg/litre (1.46 µg nickel/cm^2 per day), respectively. It appeared that multiphase microstructure was more critical to release than was the overall content of noble metals. An initial cleaning (brushing) did not change the pattern of release but did generally decrease the quantities of elements released (Wataha et al., 1992).

High-noble (Au50, Cu32, Ag12, Pd3, Zn3; atomic per cent) and noble alloys (Au36, Ag30, Cu24, Pd6, Zn3; atomic per cent) do not release detectable levels of palladium (AAS detection limit 20 µg/litre). A silver-based metal alloy (Ag55, Pd23, Cu18, Zn3; atomic per cent) released palladium at a level of 30 µg/litre (0.003 µg palladium/cm^2 per day) between 1 and 96 h (Wataha et al., 1995a).

The corrosion of a palladium alloy (Pd73, Cu14, In5; probably weight per cent) was examined after insertion in a lactic acid–saline solution by AAS and potentiodynamic measurements. This alloy had a low corrosion resistance, with a release of 0.3 µg palladium/cm^2 per day. No corresponding clinical findings in 72 patients who had partial dentures of this alloy in their mouths for up to 48 months were found (Augthun & Spiekermann, 1994).

Pfeiffer & Schwickerath (1995) determined the ion release of palladium-based dental alloys by AAS. The palladium specimens were immersed in an electrolyte (artificial saliva consisting of 0.1 mol lactic acid/litre and 0.1 mol sodium chloride/litre; pH 2.3; 37 °C) for 42 days. Ion release (µg/cm^2 per day) of the dental materials was determined as the average of the periods 1st day, 2nd to 4th day, 5th to 7th day, and 40th to 42nd day. The test solution was replaced at 3-day intervals. The examinations showed that the tested palladium–copper fusions (0.2–6 µg palladium/cm^2 per day) were less corrosion-resistant than palladium fusions with low (<3% by weight) copper contents (<0.2 µg palladium/cm^2 per day). Copper–palladium–tin fusions showed the highest palladium solubility (6–22.5 µg palladium/cm^2 per day).

The long-term corrosion behaviour of two palladium dental casting alloys was studied by Strietzel & Viohl (1992). The alloys were subjected to five simulated ceramic firings. For each alloy, three specimens were tested for 1 year (53 weeks). In weekly intervals, the artificial saliva (pH not given) was exchanged and analysed by AAS. The total palladium release after 1 year (2000 µg/cm^2 ≈ 5.5 µg/cm^2 per day) from the first alloy (Pd80, Sn6.5, Ga6.5, Cu5; weight per cent) was markedly higher than that from the second palladium–silver alloy (Pd58, Ag30, Sn6, In4; weight per cent), which released about 18 µg palladium/cm^2 (~0.05 µg/cm^2 per day).

The release of elements from several palladium-containing dental alloys into cell culture medium over 10 months was evaluated with flame AAS (detection limit 54 µg/litre). The cell culture medium was changed every month. A palladium–gold alloy (Pd48, Au35, Ga15, In2; atomic weight per cent) released an average of 0.003 µg palladium/cm^2 per day. A palladium–copper–gallium alloy (Pd74, Cu11, Ga8, In4, Sn2, Au1; atomic weight per cent) released an average of 0.005 µg palladium/cm^2 per day. A palladium–silver alloy (Pd62, Ag24, Sn9, Zn3, In2; atomic weight per cent) released 0.003 µg palladium/cm^2 per day (Wataha & Luckwood, 1998).

An *in vitro* study has measured the release of palladium from a palladium–copper–gallium alloy (Pd79.7, Ga6.0, Sn6.5, Cu5.0, Au1.0, Ru0.8; weight per cent) and gold–palladium alloy (Au51.1, Pd38.5, In9.0, Ga1.2, Ir0.2; weight per cent) with and without toothbrushing for 30 min at 200 g force. Brushing without toothpaste increased the palladium release from the silver–palladium alloy from <0.06 to 0.10 µg/cm^2 and from the palladium–copper–gallium alloy from <0.06 to 0.15 µg/cm^2. When brushing with toothpaste was done, palladium release from the gold–palladium alloy increased to 0.5 µg/cm^2 and from the paladium–copper–gallium alloy to 0.9 µg/cm^2. The alloys had a surface area of 9.08 cm^2 (Wataha et al., 1999).

Other studies by Marx (1987), Kobayashi (1989), Drápal & Pomajbík (1993), Wataha et al. (1994a), Schultz et al. (1997) and Begerow et al. (1999a) reported similar results.

5.2.4.2 Clinical studies

Salivary fluid was collected from 97 persons of both sexes in Switzerland. The palladium content in saliva was determined by AAS. Six-millilitre samples of morning saliva were collected before breakfast and before toothbrushing (Wirz et al., 1993). Three groups were formed: a control group (A) consisting of 33 healthy subjects with intact teeth, group B consisting of 32 persons with amalgam fillings and group C consisting of 32 persons with amalgam fillings and metallic dental appliances. The palladium content in saliva was higher in group B (2.8 ± 2.7 µg/litre) and significantly higher in group C (10.6 ± 7.4 µg/litre) than in control group A (1.5 ± 1.5 µg/litre).

In the same study by Wirz et al. (1993), a palladium concentration of 1400 µg/g was found in inflamed gingival tissue from a 40-year-old patient suffering from allergic reactions (mainly to nickel, chromium and fancy jewellery). The gingival sample was taken from the edge of a gold–silver–palladium crown (Wirz et al., 1993).

Augthun & Spiekermann (1994) reported that patients who had partial dentures composed of a palladium alloy (Pd73, Cu14, In5; probably weight per cent) in their mouths for up to 48 months reported no clinical symptoms.

During mechanical stimulation by continuous gum chewing, the palladium release rate from dental alloys in two patients increased strongly from 0.4 and 1.8 µg/litre saliva to 204 and 472 µg/litre saliva, respectively (Daunderer, 1993). No details about the dental status of the two patients or the time dependence of the test were given.

In a patient wearing a palladium-based crown for 0.5 years, a palladium concentration of 2.2 µg/litre saliva was determined (Kratzenstein et al., 1988). In the same article, a baseline level of <1 µg palladium/litre was reported based on results from 19 control persons.

Because of a lack of quantitative data, only a coarse estimate for a mean level of palladium intake due to iatrogenic exposure can be given. The clinical relevance of experiments with electrolytes or artificial saliva is not known for certain, because microbial processes or complexation with amino acids or peptides may not be taken into account. Variations in element release from study to study are caused by variations in the parameters of the test — for example, pH value, applied potential, composition and processing of the fusion, composition of the corrosion liquid and mechanical stress. The palladium-containing dental alloys exhibit a complex release behaviour that cannot be predicted from their nominal composition.

The results of the limited clinical studies suggest that, depending on the number of palladium-containing dental restorations and assuming a median value of 1–1.5 litre ingested saliva (Brockhaus, 1970), a daily mean intake of ≤1.5–15 µg palladium/adult per day is to be expected. It is clear from the original reports that substantial individual variation exists.

In the context of investigations on amalgam fillings, the suitability of measurements in saliva for estimating body doses has been questioned, as production and composition of saliva vary considerably. Therefore, measurements are hardly reproducible (German Federal Environment Agency, 1997).

5.3 Occupational exposure during manufacture, formulation or use

Occupational exposure occurs during mining, processing and recycling of palladium and through inhalation of palladium compounds in palladium refining and catalyst manufacture. Few data are available on concentrations at the workplace.

5.3.1 Workplace concentrations

In an early investigation (Fothergill et al., 1945), a palladium concentration of less than 5 $\mu g/m^3$ in the vicinity of various operations connected with the refining of PGMs was measured using particle filters. In the handling of dry salts, concentrations up to 32 μg palladium/m^3 were found in the atmosphere. Because of analytical shortcomings and developments in process techniques, these data are not considered reliable.

Air samples were collected in and around the mining and ore processing facilities (nickel/copper/platinum/palladium) located in Sudbury, Ontario (Canada) (Johnson et al., 1976). With the exception of the precious metals area (0.29 μg palladium/m^3), these samples did not show measurable levels of palladium (detection limit 0.003 μg palladium/m^3). The precious metals section was located within the ore processing part of the plant, and this area included the final step performed on PGM concentrates.

Air samples taken from a platinum and palladium refinery in New Jersey (USA) were reported to contain palladium at concentrations ranging between 0.001 and 0.36 $\mu g/m^3$ (Johnson et al., 1976). Weekly average palladium concentrations of 0.085 $\mu g/m^3$ and 0.028 $\mu g/m^3$ were measured in the refinery section and salts section, respectively.

Concentrations of palladium in personal air samples taken during industrial operations in which palladium compounds were handled in 1996 were reported to be between 0.4 and 11.6 µg palladium/m³, expressed as an 8-h time-weighted average concentration ($n = 4$; mean 4.81 µg/m³) (P. Linnett, personal communication, Johnson Matthey plc, May 2000).

In more recent studies, mean palladium concentrations of 3.5 µg/m³ (highest concentration 9.8 µg/m³) (with local exhaust ventilation) and 5.5 µg/m³ (highest concentration 15 µg/m³) (without dust control) were measured in the breathing zone of dental technicians (diameter of dust particles: 0.5–3 µm) (Purt, 1991). Total dust concentrations of 0.7 mg/m³ after 2 h and 1.1 mg/m³ after 6 h were determined at the workplace of dental technicians (Augthun et al., 1991). The particle size of the particulate palladium was determined using an electron microscope. No particles smaller than 6 µm in diameter were found. However, no quantitative determination of the palladium concentration was conducted, and the authors indicated that the results of their size determination were only semiquantitative.

According to the calculations of Purt (1991), a dental technician (under the most unfavourable circumstances) can take in via inhalation up to 337 mg of palladium during 20 professional years while processing palladium-based (fired) alloys (taking in a maximum of 15 µg/m³ for 45 weeks, 8 h/day, 5 days/week, for 20 years).

5.3.2 Human monitoring data

In 1974, blood samples collected from 61 platinum and palladium refinery workers in New Jersey (USA) contained no measurable amounts of palladium (<0.04 µg/100 ml) (Johnson et al., 1976). Thirty-four of the 58 workers submitting urine samples had urine palladium concentrations above the 0.21 µg/litre detection limit. The mean concentration was 1.07 µg/litre; the maximum was 7.41 µg/litre.

Forty-nine male employees in mining and ore processing at a mine in Sudbury, Ontario (Canada), were sampled twice for blood, urine and faeces, but no detectable levels were found (limits of detection: blood, 0.4 µg/kg; urine, 7 ng/litre; faeces, 20 ng/kg) (Johnson et al., 1976).

In a small human biomonitoring study, urinary platinum, palladium and gold excretion was analysed in 27 dental technicians, 17 traffic-exposed road construction workers and 17 adolescents without occupational exposure (control group) (Begerow et al., 1999b). The average urinary palladium concentrations in the control group (31.0 ± 10.8 ng/litre) and road construction workers (52.2 ± 35.3 ng/litre) correspond to the background values given in this report; in dental technicians (135.4 ± 183.5 ng/litre), however, the palladium levels were found to be significantly higher than in the control group, but not significantly higher than in the road construction workers. This would indicate that the processing of palladium-containing alloys can lead to considerable occupational exposure in dental technicians. However, it should be taken into consideration that the dental technicians had a significantly larger number of dentures, crowns and inlays containing noble metals than the other study groups (27 technicians, 246; 17 workers, 18; 17 control group, 14). The noticeable variation in the individual urinary palladium excretion within the dental technicians group (8.4–1236.2 ng palladium/litre) might therefore also reflect the palladium exposure caused by dental alloys in addition to the influence of airborne palladium. There was no significant correlation between occupational exposure to traffic exhaust and urinary palladium excretion in road construction workers.

6. KINETICS AND METABOLISM IN LABORATORY ANIMALS AND HUMANS

6.1 Absorption

Absorption of palladium appears to depend on its chemical form and the route of administration.

Generally, the absorption of metals and metal compounds is governed by their solubility in aqueous media. Interestingly, palladium dust, which is nearly insoluble in distilled water (see section 2), was found to dissolve appreciably in biological media such as gastric juice and blood serum (Roshchin et al., 1984; experimental details not given) or in aqueous solutions (under oxygen atmosphere) of biogenic compounds such as peptides and amino acids (Freiesleben et al., 1993). For example, the mass concentration of dissolved palladium at room temperature was about 40 μg/ml in an aqueous solution of the amino acid L-alanine (versus 0.3–0.4 μg/ml in water) (Freiesleben et al., 1993).

6.1.1 Absorption in animals

6.1.1.1 Salts or complexes of palladium

Palladium ions are poorly absorbed from the digestive tract. When fasted adult rats (n = 20) were given single oral doses of ^{103}PdCl$_2$ (25 μCi in 0.2 ml saline), absorption was less than 0.5% of the initial dose after 3 days. However, the amount absorbed and retained by non-fasted suckling rats (n = 15) treated similarly was significantly higher (about 5% 4 days after dosing) (Moore et al., 1974, 1975).

The same authors also studied single intravenous, intratracheal (both 25 μCi in 0.1 ml saline) and inhalation (5 μCi; aqueous aerosol; 7.2 mg/m^3; intended aerodynamic diameter 1 μm) exposures to ^{103}PdCl$_2$ in adult rats ($n \geq$ 10 per group). All three exposures resulted in a higher absorption/retention than observed for oral administration (data given in a figure only). The ranking order (see also section 6.5) was as follows: intravenous > intratracheal > inhalation > oral (suckling) > oral (adult) (Moore et al., 1974, 1975).

After topical treatment of rabbits with palladium hydrochloride (formula not specified) (see also section 6.2.1; Kolpakov et al., 1980) or of guinea-pigs with chloropalladosamine (Roshchin et al., 1984; experimental details not reported), palladium was found in all internal organs examined.

5.1.1.2 Palladium metal or metal oxides

Six months after a single intratracheal application of 50 mg palladium dust (mass median aerodynamic diameter [MMAD] 3.92 µm) to Sprague-Dawley rats, palladium particles were found to be localized intracellularly in alveolar macrophages (Augthun et al., 1991).

Quantitative data were not available.

6.1.2 Absorption in humans

There were no quantitative data available on the absorption of palladium or palladium compounds in humans.

6.2 Distribution

6.2.1 Animal studies

5.2.1.1 Distribution in organs and blood

Distribution of palladium in tissues of rats, rabbits or dogs after single oral, intravenous or intratracheal doses of palladium salts or complexes has been summarized in Table 17. The highest concentrations were found in kidney and liver, spleen, lymph nodes, adrenal gland, lung and bone. For example, 1 day after intravenous administration, palladium concentrations in liver or kidney of rats amounted to 8–21% of the administered dose. With the exception of oral treatment, palladium was detected in all tissues analysed. It could be found even 104 days after intravenous and intratracheal treatment, with maximum concentrations in spleen and lung, respectively.

There are also some data available on the distribution of palladium after repeated administration of several palladium compounds via different routes.

Table 17. Distribution of palladium in experimental animals after single-dose administration

Pd form	Route	Species	Dose[a]	Time post-exposure	Tissues under study, ranked in order of decreasing concentration	Reference
PdCl$_2$ (^{103}Pd)	oral	rat	25 µCi/rat	24 h 104 days	kidney > liver; other tissues: not detected not detected	Moore et al. (1974, 1975)
PdCl$_2$ (^{103}Pd)	intravenous	rat	25 µCi/rat	24 h 104 days	present in all tissues analysed: kidney > spleen > liver > adrenal > lung > bone > blood, heart, pancreas, fat, muscle, brain, testicle, ovary highest concentrations in spleen, kidney, liver, lung, bone	Moore et al. (1974, 1975)
PdCl$_2$ (^{103}Pd)	intravenous	rat (tumour-bearing[b])	0.16 MBq	24 h	kidney (21)[c] > liver (10) > spleen (4) > lung (2) > adrenal (1.6) > pancreas (0.4) > thymus (0.3) > cardiac muscle (0.3) > tumour[b] (0.2) > skeletal muscle (0.08) > blood (0.05) > brain (0.03)	Ando & Ando (1994)
PdCl$_2$	intravenous	rabbit	n. sp.	5 days	spleen > bone marrow > kidney > liver > lung > muscle	Orestano (1933)
PdCl$_2$ (^{103}Pd)	intratracheal	rat	25 µCi/rat	104 days	lung > kidney > spleen > bone > liver	Moore et al. (1974, 1975)
Na$_2$PdCl$_4$ (^{103}Pd)	intravenous	rat	0.5–2 µCi/rat	1, 7, 16 days	mainly in: kidney, liver, spleen	Durbin et al. (1957)
	intravenous	rat	n. sp.	1 day	liver (8.6)[c] > kidney (8.4) > muscle (1.3) > bone (1.0) > blood (0.8)	Durbin (1960)
Complexes of ^{109}Pd-haemato-porphyrin	intravenous	rabbit	1 mg/rabbit	24 h	liver, bone marrow, spleen, lymph nodes, kidney > pancreas, muscle	Fawwaz et al. (1971)

70

Table 17 (contd).

Pd form	Route	Species	Dose[a]	Time post-exposure	Tissues under study, ranked in order of decreasing concentration	Reference
Complexes of ^{109}Pd-haemato-porphyrin and proto-porphyrin	intravenous	dog	5 mg/dog	24 h	liver, lymph nodes > other organs (renal cortex, bone marrow, adrenal, spleen, duodenal mucosa, lung, renal medulla, pancreas, muscle)	Fawwaz et al. (1971)

[a] n. sp. = not specified; 1 Ci = 3.7×10^{10} Bq (dose in terms of mass n. sp.).

[b] It was found that ^{103}Pd was concentrated in viable tumour tissue but not in necrotic tumour tissue.

[c] Percentage of administered dose per gram tissue weight in parentheses.

No palladium was detected in inner organs of rabbits ($n = 10$) 21 days after intravenous injection of three doses (0.62, 0.44 and 0.18 mg/kg body weight in saline on days 0, 5 and 10) of palladium hydrochloride (formula not specified) (Kolpakov et al., 1980). However, if palladium hydrochloride was rubbed into shaved dorsal skin of rabbits ($n = 5$) at daily (6 days/week) doses of 5.4 mg/kg body weight (2% in aqueous solution), palladium was detected (spectrographically) at day 35 in blood, kidney, lung, heart, adrenal gland, brain, urinary bladder and bone (Kolpakov et al., 1980). No significant palladium levels were found in serum (and selected organs) of rabbits, guinea-pigs and mice after short-term (4 weeks) subcutaneous or intravenous administration of palladium(II) sulfate (up to 0.3 mg/animal) or palladium (up to 10 mg/animal) complexed with albumin, although palladium was present in some urine samples (Taubler, 1977).

Howarth & Cooper (1955) showed that colloidal palladium of known particle size (maximum diameter mostly 10–20 nm) was able to pass into the blood from the cerebrospinal subarachnoid space after its subarachnoid injection into cats.

Dietary administration to rats of high levels of the slightly water-soluble palladium salts palladium(II) chloride and palladium(II) sulfate (about 3000 mg palladium/kg feed for 4 weeks; corresponding to about 700 mg/kg body weight per day) resulted in palladium tissue concentrations (in mg/kg wet weight) as follows: kidney (35 and 22, respectively) > liver (2 and 3) > spleen (0.7 and 0.9) > testis (0.24 and 0.26) > blood (<0.04 and 0.16) > brain (<0.01 and <0.01). Only trace amounts of palladium (0.15 mg/kg) have been found in the kidney, and non-detectable levels in the other tissues, after administration of the water-insoluble palladium(II) oxide (about 3000 mg palladium/kg feed; corresponding to about 800 mg/kg body weight per day). The total consumption during the 4-week diet period amounted to about 2000 mg palladium/rat (i.e., 2020/1868/2276 mg palladium/rat from palladium(II) chloride/sulfate/oxide; mean body weights of the rats when started on the diets: 100–110 g) (Holbrook, 1977).

After chronic inhalation of chloropalladosamine (5.4 or 18 mg/m^3; 5 h/day, 5 days/week, for 5 months), the highest palladium concentrations (no further data given) were found (spectrographically) in lung, liver, kidney, adrenal gland and gastrointestinal tract of rats ($n = 24$) (Panova & Veselov, 1978).

Palladium dust has also been detected in alveolar macrophages (see section 6.1.1.2).

6.2.1.2 Transfer to offspring

As shown in Table 18, palladium was found not only in maternal organs but also, at considerably lower concentrations, in fetuses of rats given single intravenous doses of ^{103}PdCl$_2$ (Moore et al., 1974, 1975).

Table 18. Distribution of ^{103}Pd in maternal organs and fetuses of rats 24 h after a single intravenous injection of ^{103}PdCl$_2$ (25 µCi/rat)[a,b]

Tissue	Mean counts/g tissue
Maternal	
Kidney	588 479
Liver	319 153
Ovary	29 625
Lung	29 211
Bone	18 351
Blood	3 654
Placenta	58 321
Fetal	
Fetal liver[c]	1 429
Fetus[c]	757

[a] Adapted from Moore et al. (1975).
[b] Pregnant rats injected at 16th day of gestation (n = 13); 25 µCi = 9.25 × 10^5 Bq.
[c] Mean value from 35 fetuses (another 17 fetuses from five litters did not show significantly elevated radioactive counts).

In addition to placental transfer, palladium passage from nursing to suckling rats via milk occurred. A small amount of ^{103}Pd (10–50 counts/g tissue) was detected in tissues of young rats whose dams had received single intravenous doses of ^{103}PdCl$_2$ (25 µCi/rat) within 24 h postparturition. Twenty-five days after dosing of the dams, the bone of the sucklings had the highest radioactivity, followed by the kidney, spleen, lung and liver (Moore et al., 1974).

6.2.1.3   Subcellular distribution

The subcellular distribution of various radioactive metal ions including [103]Pd was studied in liver and tumours of intravenously dosed rats and mice (Ando & Ando, 1994). The concentrations of [103]Pd in each cellular fraction (supernatant, nuclear, mitochondrial, microsomal) of liver were relatively uniform. However, in the case of tumour tissue, most of the [103]Pd was localized in the supernatant fraction. Another study found higher concentrations of palladium in nuclei and mitochondria than in microsomes and cytosols of mouse livers after intraperitoneal administration of palladium(II) chloride (Phielepeit et al., 1989).

6.2.2 Human studies

There are a few reports on the presence of palladium in human tissues and fluids, reflecting the potential of palladium for mobility in the body.

There is a report (Daunderer, 1994) on a dental patient showing a palladium concentration of 1.8 µg/litre in saliva and of 357 µg/kg in a tooth (no further details given). Another patient had palladium concentrations of 0.9 and 1.2 µg/litre in saliva and serum, respectively (Daunderer, 1993). A palladium concentration of 3400 µg/kg has been reported in a bladder papilloma sectioned from a patient with palladium-containing dental prostheses (Daunderer, 1993). No further details were given.

For additional data on palladium concentrations in single human tissues, see sections 5.2.4 and 5.3.2.

6.3 Metabolic transformation

Certain metal ions including palladium have been reported to form stable organometallic compounds — for example, metal alkyls — in "organisms" (no details given) (Wood et al., 1978; Bonner & Parke, 1984; Morgan & Stumm, 1991). However, experimental details for palladium were not found.

For information on the biomethylation of palladium, see section 4.2.2.

6.4 Elimination and excretion

Information on the elimination and excretion of palladium is scarce. Most of the animal experiments refer to palladium(II) chloride or sodium tetrachloropalladate(II), which were found to be eliminated via faeces and urine (Table 19).

Table 19. Elimination of palladium in experimental animals following single intravenous doses

Species	Pd form	Dose per animal[a]	Time	Elimination (% of administered dose)		Reference
				Faeces	Urine	
Rabbit	$PdCl_2$	n. sp.	4 days	traces	~40	Orestano (1933)
Rat	[103]$PdCl_2$	25 µCi (0.93 MBq)	24 h	similar quantities (no details)		Moore et al. (1975)
Rat[b]	[103]$PdCl_2$	0.16 MBq	3 h	n. sp.	6.35 ± 0.52	Ando & Ando (1994)
Rat	Na_2PdCl_4 ([103]Pd)	0.5–2 µCi	4 h 7 days	n. sp. 13	60 76	Durbin et al. (1957)
Rat	Na_2PdCl_4 ([103]Pd)	n. sp.	1 day	n. sp.	74.8	Durbin (1960)

[a]   n. sp. = not specified; 1 Ci = 3.7×10^{10} Bq.
[b]   Male Donryu rats underwent subcutaneous tumour implantation in the right thigh.

After a single oral dose of [103]$PdCl_2$, a major part (>95%) of the [103]Pd was eliminated via the faeces of rats due to non-absorption (Moore et al., 1974, 1975). After intravenous administration, contents of [103]Pd were similar (Moore et al., 1974, 1975) or greater (Durbin et al., 1957) in urine than in faeces of rats. As seen in Table 19, urinary excretion rates of intravenously dosed rats and rabbits ranged from 6.4 to 76% of the administered dose. The lowest value has been observed in tumour-bearing animals (Ando & Ando, 1994).

Palladium was also detected in urine of guinea-pigs and rabbits during short-term administration of palladium(II) sulfate and a palladium–albumin complex (Taubler, 1977) or after dermal treatment (details not given) with "chloropalladium" (Roshchin et al., 1984).

Taubler (1977) found increased palladium concentrations in urine of guinea-pigs and rabbits following subcutaneous injections of palladium(II) sulfate (0.05–0.35 mg/animal; 3 times/week for 4 weeks) and in urine of guinea-pigs following exposure to palladium–albumin complex (10 mg palladium/animal; four intravenous and five subcutaneous injections over 3 weeks). A maximum palladium urine concentration of 21 mg/litre was obtained in guinea-pigs.

Palladium was also found in human urine sampled from refinery workers (see section 5.3.2). No significant increase in urinary palladium excretion was found in three volunteers who had received a high-gold dental alloy inlay and were monitored for 3 months post-insertion. This was expected, due to the very low palladium content (<0.5%) of this alloy (Begerow et al., 1999a).

6.5 Retention and turnover

Whole-body retention of [103]Pd (given as palladium(II) chloride) in adult rats (Charles River CD-1 strain) depended on the route of administration as a consequence of different absorption (see also section 6.1). Amounts retained over a period of time were lowest after single oral doses, intermediate after intratracheal and inhalation exposure and highest following intravenous administration. Retention was about 5% of the initial dose 40 days after intratracheal and 20% after intravenous exposure. Approximately 10% of the initial intravenous dose was retained at 76 days (Moore et al., 1974, 1975). From the data of Moore and co-workers, Estler (1992) estimated a biological half-life of 12 days.

In another experiment, 16 days after intravenous injection of [103]Pd (given as sodium tetrachloropalladate(II); 0.5–2 µCi/rat; $n = 3$ per group), [103]Pd could be detected in liver (1.3% of administered dose/g tissue) and kidney (0.3%/g tissue) of rats (Sprague-Dawley). Half-lives were calculated as 5 days for whole body (biphasic: 2 h, 5 days), 6 days for liver and 9 days for kidney (Durbin et al., 1957).

Mean retention values for several tissues at three time intervals (3 h, 24 h, 48 h) were determined in sarcoma-bearing (transplanted) rats (Donryu; $n = 5$ per group) injected intravenously with [103]PdCl$_2$ (Ando & Ando, 1994). Little change with time was seen for kidney (about 20% of administered dose/g tissue at each term), spleen (about 4%/g) and other tissues (muscle, pancreas, thymus, brain, bone: each <1%). The values for liver decreased slightly from 3 h to 48 h (14 to 9%/g). A marked decrease occurred in lung (8.9 to 1.8%/g), adrenal gland (2.5 to 1.1%/g) and blood (0.35 to 0.04%/g).

6.6 Reaction with body components

Like other ionic derivatives of PGMs, palladium ions are capable of binding to amino acids, proteins, DNA or other macromolecules.

Palladium(II) formed complexes with L-cysteine, L-cystine and L-methionine in solution, but not with L-histidine (Akerfeldt & Lövgren, 1964). It was also found to bind to proteins such as casein and silk protein (fibroin) (Spikes & Hodgson, 1969) or papain (Herblin & Ritt, 1964) and several enzymes, which were made inactive (see Table 26 in section 7.8). Binding may be due to interaction with sulfhydryl or other functional groups (Spikes & Hodgson, 1969). Nielson et al. (1985) found indications of binding of palladium(II) to (rat liver) metallothionein.

There was also a high affinity of palladium compounds for nucleic acids. *In vitro* studies with palladium(II) chloride and calf thymus DNA indicated that palladium(II) interacts with both the phosphate groups and bases of DNA (Pillai & Nandi, 1977). Many palladium–organic complexes were also observed to form bonds with calf thymus DNA (e.g., Jain et al., 1994) or *Escherichia coli* plasmid DNA (e.g., Matilla et al., 1994). The binding affinities of the different palladium–organic complexes varied, often parallel to their antineo-plastic activity (e.g., Lin et al., 1993). Most of the complexes appear to interact via non-covalent binding, mainly via hydrogen bonding (Mital et al., 1991; Mansuri-Torshizi et al., 1992a,b; Paul et al., 1993; Jain et al., 1994); in a few cases, however, indications for covalent binding were seen (Mansuri-Torshizi et al., 1991).

It should be noted that there are some possibilities of tuning the reactivity of palladium(II) complexes with nucleic acid components — for example, as a tool for preparing suitable anticancer drugs (Rau & van Eldik, 1996). An overview on ternary model palladium(II)–nucleic acid–protein complexes has been given by Sabat (1996).

Palladium(II) ions (potassium tetrachloropalladate(II)) were also able to form complexes with B_6 vitamins (pyridoxal, pyridoxine and pyridoxamine), which in turn reacted with nucleosides, yielding mixed-ligand complexes (Pneumatikakis et al., 1989).

7. EFFECTS ON LABORATORY MAMMALS AND *IN VITRO* TEST SYSTEMS

7.1 Single exposure

Acute toxicity data are available for several palladium compounds. LD_{50} values (Table 20) determined in rats, mice or rabbits ranged from 3 mg/kg body weight (palladium(II) chloride; rat; intravenous) to >4900 mg/kg body weight (palladium(II) oxide; rat; oral), depending on compound and route tested. Consistent with the low absorption (see section 6.1), oral administration caused the lowest toxicity. Palladium(II) chloride, potassium tetrachloropalladate(II) and ammonium tetrachloropalladate(II) had very similar intravenous LD_{50} values.

Symptoms and toxic effects observed following single exposures to several palladium forms have been compiled in Table 21. Clinical signs of toxicity included deaths, clonic and tonic convulsions, decreases in feed and water uptake, emaciation, cases of ataxia and tiptoe gait. For cardiovascular effects in rats (Wiester, 1975), the lowest effective dose after intravenous application was 0.4 mg Pd^{2+}/kg body weight. Gross pathological examination of intraperitoneally dosed rats showed prominent peritonitis with visceral adhesions (Moore et al., 1975). Histological or functional changes of the kidneys were observed with different substances (including palladium powder) and routes of application (Orestano, 1933; Fisher et al., 1975; Moore et al., 1975; Roshchin et al., 1984). Inhibition of DNA synthesis in spleen, liver, kidney and testes (Fisher et al., 1975) occurred at intraperitoneal doses of 14–56 µmol palladium(II) nitrate/kg body weight, equivalent to 1.5–6 mg Pd^{2+}/kg body weight.

Acute toxicity tests with six different palladium-containing dental alloys (pulverized; mean particle size 200 µm; palladium content: 9–80 mass per cent) have been performed with rats (Wistar, $n = 10$; 5 females, 5 males). Oral doses of 200 mg/kg body weight (in gelatin capsules) did not cause deaths or apparent toxic signs (Reuling, 1992). Histopathological changes in lung, liver and kidney of rats (Sprague-Dawley; $n = 14$) were seen 5 weeks after single oral administration of cut dental alloy material (1 g/kg body weight, in gelatin capsules; six

Table 20. Summary of LD$_{50}$ values for palladium compounds

Compound	Species (strain)	Sex; number; observation period[a]	Route[b]	LD$_{50}$ values		Reference
				mg/kg body weight[c]	mmol/kg body weight[c]	
PdCl$_2$	rat (Charles-River CD-1)	n. sp.; n = 6; n. sp.	oral	200[d]	1.13	Moore et al. (1975)
PdCl$_2$	mouse (NMRI)	n. sp.	oral	>1000	5.6	Phielepeit et al. (1989)
PdCl$_2$	rat (Charles-River CD-1)	n. sp.; n. sp.; 14 days	iv	3[e]	0.0169	Moore et al. (1975)
PdCl$_2$	rat (Charles-River CD-1)	n. sp.; n = 6; n. sp.	iv	5[d]	0.0282	Moore et al. (1975)
PdCl$_2$	rabbit	n. sp.; n = 6; n. sp.	iv	5[d]	0.0282	Moore et al. (1975)
PdCl$_2$	rat (Charles-River CD-1)	n. sp.; n = 6; n. sp.	ip	70[d]	0.395	Moore et al. (1975)
PdCl$_2$	rat (Charles-River CD-1)	n. sp.; n. sp.; 14 days	ip	123[e]	0.694	Moore et al. (1975)
PdCl$_2$	mouse (NMRI)	n. sp.	ip	87	0.492	Phielepeit et al. (1989)
PdCl$_2$	rat (Charles-River CD-1)	n. sp.; n = 6; n. sp.	itr	6[d]	0.0338	Moore et al. (1975)
PdCl$_2$	chicken		egg	>20 mg/egg	0.112 mg/egg	Ridgway & Karnofsky (1952)
PdCl$_2$·2H$_2$O	rat (Sprague-Dawley)	male; n. sp.; 14 days	oral	575	2.7[e]	Holbrook et al. (1975)
PdCl$_2$·2H$_2$O	rat (Sprague-Dawley)	male; n. sp.; 14 days	ip	85–128	0.4–0.6[e]	Holbrook et al. (1975)
PdSO$_4$	rat (Sprague-Dawley)	male; n. sp.; 14 days	oral	>1420[e]	>7	Holbrook et al. (1975)
PdSO$_4$	rat (Sprague-Dawley)	male; n. sp.; 14 days	ip	>120	>0.6[e]	Holbrook et al. (1975)

Table 20 (contd).

Compound	Species (strain)	Sex; number; observation period[a]	Route[b]	LD50 values mg/kg body weight[c]	LD50 values mmol/kg body weight[c]	Reference
PdO	rat (Sprague-Dawley)	male; n. sp.; 14 days	oral	>4900[e]	>40	Holbrook et al. (1975)
Na$_2$PdCl$_4$·3H$_2$O	mouse (ICR, Swiss)	male; n = 4 × 5; 24 h	ip	122	0.35	Jones et al. (1979)
K$_2$PdCl$_4$	rat (Charles-River CD-1)	n. sp.; n. sp.; 14 days	iv	6.4[e]	0.0196	Moore et al. (1975)
K$_2$PdCl$_4$	mouse (BALB/c)	male; n = 6 × 6; 14 days	ip	153	0.47	Williams et al. (1982)
(NH$_4$)$_2$PdCl$_4$	rat (Charles-River CD-1)	n. sp.; n. sp.; 14 days	iv	5.6[e]	0.0197	Moore et al. (1975)

[a] n. sp. = not specified.
[b] ip = intraperitoneal; itr = intratracheal; iv = intravenous.
[c] Unless otherwise specified.
[d] According to method of Deichmann & LeBlanc (1943).
[e] According to method of Litchfield & Wilcoxon (1949).

Notes:
1. Additional information is available on acute median LD$_{50}$ values in Sprague-Dawley rats (n = 10; 14 days of observation) from unpublished tests conducted according to OECD Guidelines No. 401 and 402:
   PdCl$_2$: median oral LD$_{50}$ >2000 mg/kg body weight (Johnson Matthey, 1994a);
   [Pd(NH$_3$)$_4$](HCO$_3$)$_2$: median oral LD$_{50}$ 933 mg/kg body weight (Johnson Matthey, 1995a), median dermal LD$_{50}$ >2000 mg/kg body weight (Johnson Matthey, 1997a).
2. Vehicles used for oral administration of test substances: saline (Moore et al., 1975); arachis oil (Johnson Matthey, 1994, 1995a); n. sp. (Holbrook et al., 1975; Phielepeit et al., 1989).

80

Table 21. Clinical signs and toxic effects observed after single exposures to palladium

Pd form	Animal[a]	Route	Dose[a,b]	Clinical signs and toxic effects	Reference
PdCl$_2$	rabbit	intravenous: infused over 23 min	18.6	fatal after 12 days with histopathological damage to kidney, bone marrow and liver	Orestano (1933)
		intravenous: bolus injection	0.6	rapid death, with damage chiefly to the heart	
PdCl$_2$	rabbit (n = 3)	intravenous	0.5–1.7	death: 1/3 (17th day; 1.7 mg/kg body weight dose); sluggishness, decrease in food and water uptake and in urine and faeces output	Meek et al. (1943)
PdCl$_2$	rat, rabbit (n = 6)	intravenous	n. sp.	death within 10 min (but not later during 14 days of observation); clonic and tonic convulsions; decrease in water intake and urine excretion (rats); proteinuria; ketonuria	Moore et al. (1975)
PdCl$_2$	rat, rabbit (n = 6)	intraperitoneal	n. sp.	non-survivors: necrosis of viscera survivors: peritonitis; 7% reduction in body weight; up to 80% reduction in food intake; proteinuria; ketonuria	Moore et al. (1975)
PdCl$_2$	rat (n = 24)	subcutaneous	4–24	death: 1/24 within 8 weeks survivors: no obvious impairment	Meek et al. (1943)
PdCl$_2$	mouse (NMRI) (n = 3–4)	oral	500	changes in the hepatic monooxygenase system (after 2–10 days)	Phielepeit et al. (1989)
PdCl$_2$; PdSO$_4$; Pd(NO$_3$)$_2$; K$_2$PdCl$_4$; (NH$_4$)$_2$PdCl$_4$	rat (male, S-D) (n = 12, control; n = 6–42 per compound)	intravenous: infused over 40 s	0.25–2 (of Pd^{2+})	cardiovascular effects, ventricular arrhythmias ($\geq$0.4 mg/kg body weight); at toxic doses, ventricular fibrillation and death; the simple salts are about 3 times more potent than the more complex ones	Wiester (1975)
PdCl$_2$	rat (S-D) (n = 10)	oral	2000	death (3/10 within 1–2 days), moribund (1/10 up to day 14); common signs of systemic toxicity, distended abdomen, emaciation, cases of	Johnson Matthey

Table 21 (contd).

Pd form	Animal[a]	Route	Dose[a,b]	Clinical signs and toxic effects	Reference
PdCl₂ (contd).				ataxia and tiptoe gait; in survivors and non-survivors: thickened and hardened gastric mucosa; in non-survivors: haemorrhagic lungs and abnormal colours in inner organs/tissues, ulcerated gastric mucosa	(1994a)
[Pd(NH₃)₄](HCO₃/₂	rat (S-D) (n = 5)	oral	500–2000	death (3/5, 5/5: 1000, 2000 mg/kg body weight, up to 6 days); common signs of systemic toxicity; pilo-erection, ptosis, distended abdomen, emaciation, ataxia, tiptoe gait; in survivors: sloughing of the non-glandular epithelium of the stomach; in non-survivors: haemorrhage of lungs and small intestine, abnormally coloured inner organs, thickened and hardened gastric mucosa	Johnson Matthey (1995a)
Pd(NO₃)₂	rat (male, S-D) (n = 4–7)	intraperitoneal	1.5–6 (of Pd²⁺)[c]	inhibition of DNA synthesis in spleen, liver, kidney and testis	Fisher et al. (1975)
(NH₃)₂PdCl₂[d]	rat (n = n. sp.)	inhalation (no further details)	10–688 mg/m³	no lethal effect; irritation of eyes (>50 mg/m³) and respiratory organs (>65 mg/m³); systemic effects (>82.2 ± 1.7 mg/m³ ≈ 41 mg palladium/m³): increase in concentrations of total protein, glucose and cholesterol in blood; decrease in concentrations of lactic acid and urea in blood serum	Roshchin et al. (1984)
Pd powder	rat (n = n. sp.)	enteral (intragastric) (no further details)	n. sp.	necrobiotic changes in the mucous membranes of the gastro-intestinal tract, granular dystrophy of hepatocytes, swelling in the epithelium of kidney tubules	Roshchin et al. (1984)

a   n. sp. = not specified.
b   In mg/kg body weight, unless otherwise specified.
c   Corresponding to 14–56 μmol (3.2–12.9 mg) palladium(II) nitrate/kg body weight.
d   Probably erroneously designated as [Pd(NH₃)₄]Cl₂ in the original paper.

different alloys; particle size not specified), which contained palladium at 2.5–4 weight per cent and total noble metals (gold, palladium) at 44–62%. However, alloys of a higher noble metal (gold, palladium, platinum) content (78–97%) containing palladium at 4–26.5% did not elicit significant effects (Culliton et al., 1981).

7.2 Short-term exposure

Most of the short-term studies with pure palladium compounds are older studies not including histopathological examinations.

Effects recorded in rodents and rabbits after short-term (repeated) exposure to various palladium forms by different routes and dosage regimens are summarized in Table 22. They include sluggishness, weight loss, exudation of mucus and pus (in rabbits following dermal application of 5.4 mg palladium hydrochloride [formula not specified]/kg body weight; Kolpakov et al., 1980), haematoma (in guinea-pigs after intravenous injection of 0.1–1 mg palladium(II) sulfate/kg body weight; Taubler, 1977) and changes in parameters of drug-metabolizing systems in rat liver (after administration of 230 mg palladium(II) chloride/kg body weight or 360 mg palladium(II) oxide/kg body weight in diet or of a saturated solution of palladium(II) chloride ($PdCl_2 \cdot 2H_2O$) given as drinking fluid or after intraperitoneal dosage of 12.9 mg palladium(II) nitrate/kg body weight; Holbrook et al., 1976). Only one of the compounds tested caused deaths (sodium tetrachloro-palladate(II)–egg albumin in mice after intravenous doses of 2.7 mg/kg body weight and more; Taubler, 1977).

A more recent unpublished 28-day toxicity study performed according to EEC Method B7 (described in Commission Directive 92/69/EEC) is available for tetraammine palladium hydrogen carbonate (Johnson Matthey, 1997b). Oral administration of this compound to rats (Sprague-Dawley Crl:CD BR strain), by gavage (with distilled water as a vehicle), at dose levels of 1.5, 15 or 150 mg/kg body weight per day for 28 consecutive days produced treatment-related changes mainly at the two higher concentrations (see Table 22). The effects seen at 15 mg/kg body weight per day were confined to histopathological changes, namely reduced levels of haemosiderin in spleen and increased numbers of mucus-secreting goblet cells in the glandular gastric mucosa. The no-observed-adverse-effect level (NOAEL) was

Table 22. Short-term toxicity of palladium

Pd form	Species[a]	Route/dosage regimen	Dose (mg/kg body weight per day)	Effects	Reference
$PdCl_2$	rat (S-D) ($n = 4$–8)	in diet (feed) 13–30 mmol/kg feed over 4 weeks	230–530[b]	no change (high dose) or increase (low dose) in activity of several hepatic microsomal enzymes	Holbrook et al. (1976)
$PdCl_2 \cdot 2H_2O$	rat (S-D) ($n = 8$)	in diet (drinking fluid) saturated aqueous solution over 8 days	–	no statistically significant changes in body or relative organ weights; decrease in activity of hepatic microsomal enzymes	Holbrook et al. (1975, 1976)
$Pd(NO_3)_2$	rat (S-D) ($n = 4$)	intraperitoneal 14–113 µmol/kg body weight per day for 2 days; measure: 3rd day	3.2–25.9	increase in hexobarbital-induced sleeping time by about 60% from 56 µmol (12.9 mg $Pd(NO_3)_2$) ($\approx$6 mg Pd)/kg; decrease in hepatic microsomal enzyme levels	Holbrook et al. (1976)
$PdSO_4$	guinea-pig ($n = 15$)	intravenous 0.05–0.35 mg per animal; 5 injections over 2 weeks	0.1–1[c]	collapse of the dorsal sapheous vein; extreme haematoma and necrosis in leg (injection site)	Taubler (1977)
$PdSO_4$	rat (S-D) ($n = 4$)	in diet (drinking fluid or feed) saturated aqueous solution or 29.8 mmol/kg feed for 1, 4 weeks	– 600[d]	no effect on activity of hepatic microsomal enzymes	Holbrook et al. (1976)
PdO	rat (S-D) ($n = 4$)	in diet (feed) 29.8 mmol/kg for 4 weeks	360[e]	no effect on activity of hepatic microsomal enzymes; decrease in yield of microsomal protein	Holbrook et al. (1976)

Table 22 (contd).

Pd form	Species[a]	Route/dosage regimen	Dose (mg/kg body weight per day)	Effects	Reference
K_2PdCl_4	rat (Charles River; CD-1) (n = 10)	in diet (drinking-water) 92 and 184 ppm (mg/litre) for 33 days	13, 26[f]	no abnormalities in general appearance, body weights or urinalysis	Moore et al. (1975)
$[Pd(NH_3)_4](HCO_3)_2$	rat (S-D) (n = 5)	oral (gavage) 1.5–150 mg/kg body weight per day for 28 days	1.5, 15, 150	reduced body weight gain, anaemia, increase in absolute and relative kidney weight (150); histopathological changes in liver and kidney (150); increase in absolute stomach (150, 15); increase in absolute brain and ovary weight in females (15, 1.5)	Johnson Matthey (1997b)
Pd hydrochloride[g]	rabbit (n = 5)	dermal 5.4 mg/kg body weight per day (as 2% solvent on shaved dorsal skin) over 56 days (excluding Sundays)	5.4	from 19th day: sluggishness; weight loss (18%); exudation of mucus and pus from nose	Kolpakov et al. (1980)
Na_2PdCl_4–egg albumin	mouse (n = 15)	intravenous 0.08, 0.25, 2.5 mg/mouse; 3 times/week for 3 weeks	2.7–83[h]	death (7/15; 10/15; 13/15)	Taubler (1977)
Colloidal Pd	rabbit (n = 2)	subcutaneous 5 mg 1% solution/day (= 0.05 mg/day) over 2 months	–	no symptoms	Kauffmann (1913)

[a] n = number per group; S-D = Sprague-Dawley.
[b] Calculated for young rats, using a factor of 0.1 (2300–5310 mg/kg feed).

Table 22 (contd).

^{c} Calculation: 300–350 g body weight.
^{d} Calculated with a factor of 0.1 (6020 mg/kg feed).
^{e} Calculated with a factor of 0.1 (3636 mg/kg feed).
^{f} Calculation: 50 ml/day water intake, 350 g body weight.
^{g} Formula not specified.
^{h} Calculation: 30 g body weight.

considered by the authors to be 1.5 mg/kg body weight per day, although significant increases in absolute brain and ovary weight have been observed in females of this dose group.

Effects observed in studies with palladium-containing dental alloy material are compiled in the following text.

Six different pulverized palladium-containing dental alloys have been tested for their subacute toxicity in rats (Wistar, $n = 10$, 5 females, 5 males). Compared with controls, all specimens caused significant histopathological changes in lung, liver, kidney, small intestine and colon following daily oral doses of 1000 mg/kg body weight for 7 days (and sacrifice after 7 additional days). The strongest effects were seen with compositions containing copper and/or indium: Pd73, Cu14, In5; Pd74, Sn16, Cu9; and Au51, Pd39, In4–9. The weakest effects were observed for the Pd58, Ag30, Sn6 alloy, and the changes due to the Pd80, Sn7, Ga6–8 and Au77, Pt10, Pd9 alloys fell in between (Fuhrmann, 1992; Reuling, 1992; Reuling et al., 1992).

Some histocompatibility tests have been performed with several implants of palladium-containing dental alloys of different composi-tions. Four to 12 weeks after subcutaneous implantation of the experi-mental alloys into guinea-pigs (Niemi & Hensten-Pettersen, 1985; Bessing & Kallus, 1987; Reuling, 1992), mice (Eisenring et al., 1986) or rats (Kansu & Aydin, 1996) or after intramuscular implantation into rabbits (Reuling, 1992), the tissue reactions at the implantation site varied from slight to severe. The most extreme reactions were recorded when copper and/or indium were present as additional components (Niemi & Hensten-Pettersen, 1985; Bessing & Kallus, 1987; Reuling, 1992; Kansu & Aydin, 1996). Altogether, the contribution of palladium to adverse effects of dental mixtures is not clear.

7.3 Long-term exposure

None of the long-term studies was performed according to current toxicological guidelines. Some of the available studies are documented in insufficient detail or performed without histopathological examina-tions.

Mice given palladium(II) chloride in drinking-water (5 mg palladium/litre) from weaning until natural death showed suppression of body weight gain together with longer life span (in males, but not in females; mean age of palladium-fed males 555 days versus 444 days in controls) and an increase in amyloidosis of several inner organs. There was also an approximate doubling (27.7%, n = 18/65, versus 13.8% in controls, n = 11/80) of malignant tumours (Schroeder & Mitchener, 1971; see also section 7.7).

The complex ammine salt chloropalladosamine was tested in rats for its "chronic" toxicity after inhalative exposure (Panova & Veselov, 1978). Rats (n = 24 per group) exposed to 5.4 and 18 mg chloropalladosamine/m^3 (5 h/day, 5 days/week, for 5 months; 30 days recovery; analytical control of exposure concentrations; particle size not specified) showed slight temporary (low dose) or significant permanent (high dose) changes in several blood serum and urine parameters, indicating damage to function of liver and kidney. Additionally, sluggishness, reduced body weight gain and changes in organ weights (increase: kidney, heart; decrease: lung, liver) were observed. First changes began after 2 weeks. Additionally, a review article by Roshchin et al. (1984) summarized toxicity data from rats exposed to this salt enterally (for about 6 months). However, experimental details are not specified exactly, and chloropalladosamine was (erroneously) also designated as tetraammine palladium(II) chloride ([Pd(NH$_3$)$_4$]Cl$_2$). For enteral exposures, doses of 0.08 mg/kg body weight and 8 mg/kg body weight were considered to be the no-observed-effect level (NOEL) and the toxic concentration, respectively. Observed effects included reduction in body weight, decreased haemoglobin content in peripheral blood, changes in several blood serum parameters, indicating metabolic disturbances, changes in activity of several enzymes, and functional and morphological (membranous glomerulonephritis) changes in the kidney. (The latter study was not taken into consideration for the evaluation because of the unusual route of exposure [enteral] and lack of histological information.)

There are two studies (Roshchin et al., 1984; Augthun et al., 1991) available investigating the chronic effects of palladium dust. Daily oral administration of 50 mg palladium powder/kg body weight to rats (strain, sex and number not specified) for 6 months resulted in delayed body weight gain, shortening of the prothrombin clotting time, a decrease in urea and lipoprotein contents and an increase in albumin concentrations in blood serum, and a decreased density of urine (Roshchin et al., 1984). In the second study, 6 months after a single intratracheal application of 50 mg palladium dust (~143 mg/kg body weight), the lungs of rats (Sprague-Dawley, female, $n = 10$) were histopathologically examined. There were several signs of inflammatory responses (peribronchial inflammation, lymphocyte infiltration, interstitial pneumonia, formation of granulomata) observed, but no indications of interstitial fibrosis or carcinogenic changes (Augthun et al., 1991).

Some long-term implantation studies with palladium-containing alloys have been performed. A silver–palladium–gold dental alloy implanted subcutaneously for 504 days in rats was found to cause tumours at the implantation site (see section 7.7).

7.4 Irritation and sensitization

7.4.1 Skin irritation

Several palladium compounds are capable of eliciting dermal irritation. A series of eight palladium compounds was tested with male albino rabbits ($n = 6$ per group), according to US National Institute of Occupational Safety and Health procedures and evaluation criteria (adapted from those of US FDA, 1973). Twenty-four and 72 h after application of the substances (0.1 g plus 0.1 ml water) onto intact and abraded dorsolateral skin, skin irritation was noted. Evaluation on the basis of combined worst intact and abraded scores resulted in the following ranking (in decreasing order of severity): $(NH_4)_2PdCl_6$ > $(NH_4)_2PdCl_4$ > $(C_3H_5PdCl)_2$ (allyl palladium chloride dimer) > K_2PdCl_6 > K_2PdCl_4 > $PdCl_2$ > $(NH_3)_2PdCl_2$ > PdO. The first three compounds (causing erythema, oedema or eschar) were considered as unsafe for skin contact; only the last two turned out to be safe. The middle three were non-irritant to intact skin but caused erythema in abraded skin (Campbell et al., 1975). The degree of irritation may correspond to the solubility of these compounds (see section 2.2.2). In another study, palladium hydrochloride (formula not provided) was applied (5.4 mg/kg body weight per day; 2% aqueous solution) to shaved dorsal skin of rabbits ($n = 5$) over 8 weeks. Dermatitis was observed beginning on day 7 (Kolpakov et al., 1980). Palladium(II) chloride tested on intact skin of rabbits ($n = 3$), according to Guideline No. 404 of the Organisation for Economic Co-operation and Development

(OECD), produced a primary irritation after 1–72 h of observation and was classified as a moderate irritant to rabbit skin according to the Draize classification scheme. Skin reactions 7 days after treatment were crust formation and desquamation (Johnson Matthey, 1994b). No dermal reactions have been found with tetraammine palladium hydrogen carbonate applied to intact skin of rabbits according to the same protocol (Johnson Matthey, 1995b).

7.4.2 Eye irritation

Palladium(II) chloride (10 mg deposited on the eye surface) caused corrosive conjunctival lesions and severe inflammation of the cornea and anterior chamber of the eyes of rabbits ($n = 6$). These effects were observed at 24 h and persisted at 48 and 72 h. At the same dose, no reaction was observed with palladium(II) oxide, and no reaction was noted with either platinum oxide or platinum dichloride, tested according to the same protocol (Hysell et al., 1974). A single application (according to OECD Guideline No. 405) of tetraammine palladium hydrogen carbonate to the non-irrigated eye of one rabbit produced severe lesions in the cornea, conjunctiva and nictitating membrane within 24 h. The substance was classified (according to a modified Kay and Calandra classification system) as at least a very severe irritant to the rabbit eye (Johnson Matthey, 1995c).

After inhalation exposure of rats to chloropalladosamine (formula in the paper erroneously given as $[Pd(NH_3)_4]Cl_2$, or tetraammine palladium(II) chloride), mucous membranes of the eyes were affected (signs of conjunctivitis or keratoconjunctivitis, depending on concentration; details of exposure regimen not given). The threshold concentration was given as 50 mg/m³ (Roshchin et al., 1984).

7.4.3 Sensitization

Some palladium compounds have been found to be potent sensitizers of the skin (Table 23). Palladium(II) chloride was a stronger sensitizer than the well known potent sensitizer nickel sulfate in the guinea-pig maximization test (Wahlberg & Boman, 1990). Besides palladium(II) chloride (Wahlberg & Boman, 1992), tetraammine palladium hydrogen carbonate also produced a 100% sensitization rate in the guinea-pig maximization test (Johnson Matthey, 1997c). Significant primary immune responses have been obtained with palladium(II)

Table 23. Summary of sensitization tests and allergic reactions of palladium compounds

Pd compound	Species	Procedure; route[a]	Result	Reference
PdCl$_2$	guinea-pig	GPMT pretreatment: FCA induction[b]: intra- and epidermal (0.03 and 2.5% in water) challenge: epidermal (0.63–1.25% in saline)	sensitization: +	Boman & Wahlberg (1990); Wahlberg & Boman (1990, 1992)
	guinea-pig	H & S pretreatment: FCA induction: intradermal (0.03% in saline), challenge: epidermal (0.3–2.5% in saline)	sensitization: +	Wahlberg & Boman (1992)
	mouse (BALB/c)	PLNA	sensitization: +	Schuppe et al. (1998)
[Pd(NH$_3$)$_4$](HCO$_3$)$_2$	guinea-pig	GPMT induction: intra- and epidermal (0.01 and 5% in arachis oil) challenge: epidermal (1–2% in arachis oil)	sensitization: +	Johnson Matthey (1997c)
PdSO$_4$	rabbit, guinea-pig, mouse	induction: subcutaneous or intravenous (up to 0.35 mg/animal; 3 times/week for 3–4 weeks) challenge: intradermal (up to 0.03 mg/animal, 10–14 days post-exposure)	sensitization: −	Taubler (1977)
Pd hydrochloride[c]	rabbit	induction: intravenous (3 times into ear, 0.18; 0.44; 0.62 mg/kg body weight at 5-day intervals) challenge: 1% (shaved skin)	sensitization: + (confirmed by lymphocyte test)	Kolpakov et al. (1980)

Table 23 (contd).

Pd compound	Species	Procedure; route[a]	Result	Reference
Na_2PdCl_4 $(NH_4)_2PdCl_6$	mouse (BALB/c)	PLNA	sensitization: +	Schuppe et al. (1998)
Na_2PdCl_4 –egg albumin	rabbit, mouse	induction: intravenous (up to 10 mg Pd/animal, 3 times/week for 3 weeks) challenge: intradermal	sensitization: –	Taubler (1977)
Na_2PdCl_4 complexed to egg albumin	guinea-pig	induction: subcutaneous (3 times/week for 3 weeks) challenge: intradermal (including passive transfer, spleen cells)	sensitization: +	Taubler (1977)
Na_2PdCl_4 complexed to guinea-pig albumin	guinea-pig	induction: subcutaneous (3 times/week for 3 weeks) challenge: intradermal (including passive transfer, spleen cells)	sensitization: +	Taubler (1977)
$(NH_4)_2PdCl_4$	rat	intravenous (repeated injections of 0.1 mg/kg body weight for 3 weeks)	no stimulation of IgE production	Murdoch & Pepys (1986)
Several Pd salts: $PdXCl$, PdX_2,[d] K_2PdCl_4, $K_2Pd(NO_2)_4$, $K_2Pd(SCN)_4$, $Pd(NH_3)_2Cl_2$ $Pd(NH_3)_2(NO_2)_2$	cat	intravenous	bronchospasms; increase in serum histamine	Tomilets & Zakharova (1979)

Table 23 (contd).

Pd compound	Species	Procedure; route[a]	Result	Reference
K_2PdCl_4:	guinea-pig	intradermal into ear induction: 65 µg/animal (~0.08 mg/kg body weight) challenge: 163 µg/animal (~0.19 mg/kg body weight)	sensitization: +	Tomilets & Zakharova (1979)
+ ovalbumin			anaphylactic shock increase in serum histamine	
− ovalbumin				

[a] FCA = Freund complete adjuvant; GPMT = guinea-pig maximization test; PLNA = popliteal lymph node assay, measuring the local primary immune response after single subcutaneous injection of soluble test compound; H & S = method developed by I.M. van Hoogstraten & R.J. Scheper (1990, personal communication cited in Wahlberg & Boman, 1992); similar to GPMT (Note: The H & S-method was used because the GPMT fails to sensitize a sufficient number of animals with nickel sulfate).

[b] Induction = induction of hypersensitivity (for details/terminology, see Magnusson & Kligman, 1970; OECD, 1992; IPCS, 1999).

[c] Formula not specified.

[d] X = 5-sulfo-8-mercaptochinolinate.

chloride and sodium tetrachloropalladate(II), potassium tetrachloro-palladate(II) and ammonium hexachloropalladate(IV) in the popliteal lymph node assay in BALB/c mice. Single subcutaneous injections of these compounds induced dose-dependent reactions, with popliteal lymph node cellularity increasing up to 8-fold (Kulig et al., 1995; Schuppe et al., 1998). Topical application of potassium tetrachloro-palladate(II) in dimethyl sulfoxide on both ears of mice on 4 con-secutive days induced a significant local response in auricular lymph nodes (Kulig et al., 1995; Schuppe et al., 1998). Additionally, potent T-cell sensitizing properties have been demonstrated using an adoptive popliteal lymph node assay. If splenic T cells from mice treated intra-peritoneally or intranasally with one of the palladium halide salts for 6–12 weeks were transferred to syngeneic recipients, they responded, after local restimulation, to all three palladium salts (Kulig et al., 1995; Schuppe et al., 1998).

In one study with intravenous administration of several palladium salts, signs of respiratory sensitization were reported (see Table 23; Tomilets & Zakharova, 1979).

There are some animal model studies using the guinea-pig max-imization test method or a similar method (Table 24) to determine if the multiple sensitivity to palladium and other metals sometimes observed in humans (see chapter 8) is due to cross-reactivity or to multiple sensitization. In guinea-pigs sensitized to palladium(II) chlor-ide, cross-reactivity to nickel sulfate was observed. However, animals induced with nickel sulfate or chromate or cobalt salts did not react upon challenge with palladium(II) chloride.

Somewhat divergent results have been obtained in tests studying cross-reactivity between palladium and nickel by repeated open appli-cations to the skin of guinea-pigs. In these experiments, animals were induced with palladium(II) chloride ($n = 27$) or nickel sulfate ($n = 30$) according to the guinea-pig maximization test method and then treated once daily for 10 days according to the repeated open application test by applying the inducing allergen (palladium(II) chloride or nickel sulfate) as well as the possibly cross-reactive compound (nickel sulfate or palladium(II) chloride) and the vehicle topically in guinea-pigs. While in the previous study (see Table 24, Wahlberg & Boman, 1992), 87% of guinea-pigs (13/15) sensitized with nickel sulfate reacted to nickel sulfate and none reacted to palladium(II) chloride, a similar

Table 24. Tests of cross-reactivity between palladium and other metal compounds in guinea-pigs

Inducing substance[a,b]	Test[c]	Challenge substance	Result[d-f]	Reference
K_2CrO_4 (intradermal/topical, 0.5% in water/1% in petrolatum)	GPMT ($n = 15$)	$PdCl_2$ (1% in petrolatum)	– (0/15)	Liden & Wahlberg (1994)
$CoCl_2$ (intradermal/topical, 1% in water/5% in petrolatum)	GPMT ($n = 15$)		– (0/15)	Liden & Wahlberg (1994)
$NiSO_4$ (intradermal, 0.3% in saline)	H & S ($n = 15$)	$PdCl_2$ (0.625–2.5% in saline)	– (0/15)	Wahlberg & Boman (1992)
$PdCl_2$ (intradermal/epidermal, 0.03%/2.5% in water)	GPMT ($n = 10$)	$NiSO_4$ (0.5% in saline)	+ (8/10)	Wahlberg & Boman (1992)
$PdCl_2$ (intradermal, 0.03% in saline)	H & S ($n = 15$)	$NiSO_4$ (0.5% in saline)	(+) (2/15)	Wahlberg & Boman (1992)

[a] Induction = induction of hypersensitivity (for details/terminology, see Magnusson & Kligman, 1970; OECD, 1992; IPCS, 1999).
[b] Purity of $PdCl_2$: <20 µg nickel/g; purity of $NiSO_4$: <1 µg palladium/g.
[c] GPMT = guinea-pig maximization test; H & S = method developed by I.M. van Hoogstraten & R.J. Scheper (1990, personal communication cited in Wahlberg & Boman, 1992); similar to GPMT (Note: The H & S-method was used because the GPMT fails to sensitize a sufficient number of animals with nickel sulfate).
[d] + = sensitization observed; – = sensitization not observed; (+) = sensitization observed, but statistically not significant.
[e] Number positive/number tested per concentration (induction and challenge at 48 h post-challenge) in parentheses.
[f] Results for the positive controls (induction and challenge by the same substance) in the GPMT were as follows: $PdCl_2$ (10/10); K_2CrO_4 (8/15); $CoCl_2$ (5–11/15), positive controls for the H & S-method: $PdCl_2$ (5/15); $NiSO_4$ (12–13/15).

95

reactivity has been found for palladium(II) chloride (23%) and nickel sulfate (30%) in repeated open application testing. There was also a difference in reactivity to nickel sulfate in animals sensitized with palladium(II) chloride (80% in the previous study versus only 7% in repeated open application tests). However, consistent results have been found from both tests for the reactivity of palladium(II) chloride in palladium-sensitized animals (both 100%). The concordance between the outcome of the repeated open application tests and results of patch tests carried out before repeated open application testing was high for palladium(II) chloride (100%) and low for nickel sulfate (10–40%). Additionally, it has been found that topical treatments with palladium(II) chloride alone can induce sensitivity (e.g., in control animals without preceding intradermal administration of palladium(II) chloride). So it remains unclear if reactivity to palladium(II) chloride in animals sensitized with nickel sulfate is due to cross-reactivity or to the induction of sensitivity by the repeated treatments. On the other hand, reactivity to nickel sulfate in animals sensitized with palladium(II) chloride could be attributed to cross-reactivity (Wahlberg & Liden, 1999).

Rats (Hooded Lister) treated with sera from rats induced with ammonium tetrachloroplatinate(II) ($(NH_4)_2PtCl_4$) in its conjugated form with ovalbumin did not react with the corresponding palladium salt conjugate in the passive cutaneous anaphylaxis test. This was confirmed negative by the radioallergosorbent test (Murdoch & Pepys, 1985).

Preliminary *in vitro* experiments with T-cell hybridomas showed that palladium(II) and palladium(IV) compounds may be able to induce autoimmunity by presentation of cryptic self-peptides. It was proposed that such altered peptides could be generated via formation of metal–protein complexes or, in the case of palladium(IV), which has a high oxidizing capacity, also via oxidation processes resulting in denaturation of proteins (Griem & Gleichmann, 1995). There were also preliminary indications that palladium(IV) was able to alter subcellular localization of scleroderma autoantigen fibrillarin in mouse epithelial cells and macrophages (Chen et al., 1998).

Studies of palladium sensitization in humans are reported in sections 8.1 and 8.2.

7.5 Reproductive and developmental toxicity

Only few data are available on reproductive and developmental toxicity; none of the studies was performed according to current guide-lines.

Male mice ($n = 3$) administered a total subcutaneous dose of 0.02 mmol palladium(II) chloride (~3.54 mg = 2.1 mg Pd^{2+})/kg body weight in 30 daily doses showed a reduction in mean testis weight. A single subcutaneous injection of 0.02 mmol/kg body weight had no effect on the weight of the testis in rats. The same single dose given to male rats ($n = 3$) intratesticularly caused a decrease in mean testis weight, total necrosis of testis and destruction of all spermatozoa within 2 days (Kamboj & Kar, 1964).

A single intraperitoneal dose of palladium(II) nitrate (e.g., 1.5 mg Pd^{2+}/kg body weight) resulted in inhibition of DNA synthesis in testis of rats as well as in other organs (Fisher et al., 1975; see also Table 21).

Four-day-old chicken embryos (6–10 developing eggs) injected with several dose levels of palladium(II) chloride ($PdCl_2 \cdot 2H_2O$) above and below the estimated LD_{50} (>20 mg/egg) and examined grossly for abnormalities (until the 18th day of incubation) showed no apparent signs of teratogenicity (Ridgway & Karnofsky, 1952).

7.6 DNA interactions and mutagenicity

7.6.1 Interaction with DNA

Palladium compounds have been found to interact with isolated eukaryotic and plasmid DNA *in vitro* (see section 6.6). This interaction induced conformational changes in DNA structure, as seen in *in vitro* studies with calf thymus DNA (Shishniashvili et al., 1971; Pillai & Nandi, 1977) and pUC8 plasmid DNA from *Escherichia coli* (Navarro-Ranninger et al., 1992, 1993; Matilla et al., 1994). Palladium compounds tested included palladium(II) chloride and inorganic and organic palladium(II) complexes. Some cyclometallated palladium complexes can lead to base-selective DNA cleavage (Suggs et al., 1989). Studies using Fenton systems showed that palladium ions (derived from palladium(II) chloride plus hydrogen chloride, thus forming hydrogen tetrachloropalladate(II)) were able to enhance hydroxyl radical-mediated DNA damage, e.g., strand breakage of supercoiled pBR322 DNA, due to promotion of hydroxyl radical production (Liu et al., 1997). It is not clear whether the *in vitro* reactions described above also apply to cellular systems.

7.6.2   *Mutagenicity*

Several mutagenicity tests of different palladium compounds with bacterial or mammalian cells (Ames test with *Salmonella typhimurium*; SOS chromotest with *Escherichia coli*; micronucleus test with human lymphocytes) *in vitro* gave negative results (Table 25). As an exception, tetraammine palladium hydrogen carbonate induced a clastogenic response to human lymphocytes *in vitro* (see Table 25).

An *in vivo* genotoxicity test (micronucleus test in the mouse) has been performed with tetraammine palladium hydrogen carbonate in compliance with OECD Guideline No. 474. Single oral doses ranging from 125 to 500 mg/kg body weight did not produce positive results (Johnson Matthey, 1998).

7.7   Carcinogenicity

There is only little information available on the carcinogenic potential of palladium.

Mice (white, Swiss Charles River CD) given palladium(II) chloride at 5 mg Pd^{2+}/litre in drinking-water (corresponding to about 1.2 mg Pd^{2+}/kg body weight per day by assuming a body weight of 0.03 kg and a daily water uptake of 7 ml) over a lifetime (from weaning to natural death) developed tumours in both sexes (19/65 [29%] versus 13/80 [16%] in control; sex-related distribution not specified). Only one tumour found in the exposed group was not malignant. Most of the malignant tumours were either lymphoma–leukaemia types (10 versus 2 in controls) or adenocarcinoma of the lung (6 versus 1 in controls). The increase in malignant tumours (18/65 [27.7%] versus 11/80 [13.8%]) was statistically significant ($P < 0.05$) compared with the simultaneous control group. This was not the case in comparison with another smaller, non-simultaneous control group, which had differing rates of malignant (6/41 [14.6%]) and total (11/41 [26.8%]) tumours

Table 25. *In vitro* mutagenicity tests with palladium compounds

Compound	Assay[a]	Species	Dose range	Result	Reference
K_2PdCl_4 K_2PdCl_6 K_2PdBr_6 $[Pd(NH_3)_4]Cl_2$	Ames (MA n. sp.)	*Salmonella typhimurium* TA98, TA100	0.1–1000 µg/plate	–	Suraikina et al. (1979)
K_2PdBr_4 $K_2Pd(NO_2)_4$ *cis-/trans-*$(NH_3)_2PdCl_2$ $K_2Pd(SCN_4)$	Ames (MA n. sp.)	*Salmonella typhimurium* TA98, TA100, TA1535, TA1538	0.1–1000 µg/plate	–	Suraikina et al. (1979)
K_2PdCl_4	Ames (+/– MA)	*Salmonella typhimurium* TA98, TA100	0.8–100 nmol/plate	–	Uno & Morita (1993)
K_2PdCl_6 $(NH_4)_2PdCl_4$ $(NH_4)_2PdCl_6$	Ames (+/– MA)	*Salmonella typhimurium* TA97a, TA98, TA100, TA102	5–500 µg/plate	–	Bünger et al. (1996); Bünger (1997)
$[Pd(NH_3)_4](HCO_3)_2$	Ames (+/– MA)	*Salmonella typhimurium* TA98, TA100, TA1535, TA1537, TA1538	0.15–500 µg/plate	–	Johnson Matthey (1995d)
$PdCl_2$ K_2PdCl_4 $Pd(NH_3)_2I_2$ $Pd(NH_3)_2Cl_2$ $Pd(NH_3)_4Cl_2$	SOS chromotest (MA n. sp.)	*Escherichia coli* PQ37	3–1147 µmol	–	Gebel et al. (1997); Lantzsch & Gebel (1997)

99

Table 25 (contd).

Compound	Assay[a]	Species	Dose range	Result	Reference
$PdCl_2$ K_2PdCl_4 $Pd(NH_3)_2I_2$ $Pd(NH_3)_2Cl_2$ $Pd(NH_3)_4Cl_2$	micronucleus test (MA n. sp.)	human peripheral lymphocytes	100–600 µmol	–	Gebel et al. (1997)
$[Pd(NH_3)_4](HCO_3)_2$	chromosome aberration test (+/– MA)	human lymphocytes	23–555 µg/ml	$(+)^b$	Johnson Matthey (1997d)

[a] MA = metabolic activation; n. sp. = not specified.
[b] Induction of statistically significant non-dose-related increases in the frequency of cells with chromosome aberrations at one dose level (370 µg/ml; + MA only).

(Schroeder & Mitchener, 1971). The increased tumour rate might be caused by a significantly enhanced longevity (mean age of the last 10% of surviving males: 815 ± 27.1 versus 696 ± 19.2 days in controls). Other limitations of this study refer to dosage regimen (only one dose was tested) and protocol (tumour rates were pooled for males and females).

Subcutaneous implantation of a silver–palladium–gold alloy led to formation of tumours (fibrosarcoma, myosarcoma, fibroma and fibroadenoma) at the implantation site in 7 of 14 rats after 504 days. However, it was not clear if the observed carcinogenicity was due to the chemical components or to the chronic physical stimulus of the imbedded alloy (Fujita, 1971).

No studies on carcinogenicity after inhalation exposure to palladium were available.

7.8 Effects on cellular functions

7.8.1 *Miscellaneous cytotoxic effects*

There are numerous *in vitro* studies on the cytotoxic effects of palladium(II) chloride (Hussain et al., 1977; Nordlind, 1986; Clothier et al., 1988; Wataha et al., 1991b, 1995b; Nordlind & Liden, 1993; Schedle et al., 1995; Schmalz et al., 1997a), palladium(II) sulfate (Rapaka et al., 1976), several inorganic/organic palladium complexes (Kolesova et al., 1979; Aresta et al., 1982; Mital et al., 1992; Bünger et al., 1996; Bünger, 1997) and an undefined Pd^{2+} compound (Eimerl & Schramm, 1993; Chiu & Liu, 1997). These studies indicate that palladium ions are capable of inhibiting most major cellular functions. Cytopathogenic effects observed included inhibition of DNA synthesis in mouse or human cell lines (Nordlind, 1986; Wataha et al., 1991b; Nordlind & Liden, 1993; Schedle et al., 1995), inhibition of RNA synthesis in purified rat liver nuclei (Mital et al., 1992), inhibition of protein synthesis or decrease in total protein content or mitochondrial activity in mouse, hamster or human cells (Clothier et al., 1988; Wataha et al., 1991b, 1995b; Schmalz et al., 1997a), necrosis in mouse and human fibroblasts (Schedle et al., 1995), reduction of cell viability in mouse and human cell lines (Bünger et al., 1996; Bünger, 1997), loss of membrane integrity in mouse and human macrophages (Wataha et al., 1995b), inhibition of (rat peritoneal) macrophage chemotaxis

(Aresta et al., 1982) and potentiation or inhibition (depending on concentration) of glutamate toxicity in rat cerebellar granule cells (Eimerl & Schramm, 1993). Tests using bovine heart tissue cultures and submitochondrial particles showed inhibition of respiration and ATPase activity (Kolesova et al., 1979); changes in collagen synthesis processes could be induced in lung organ cultures from neonatal rats (Hussain et al., 1977) or in chick embryo cartilage explant (Rapaka et al., 1976).

Altogether, the rates of response depend on test system and parameter tested. For example, human macrophages are less sensitive to palladium(II) chloride than mouse fibroblasts (Wataha et al., 1995b). With regard to parameters, the following order of *in vitro* responses was found: DNA synthesis decreased most rapidly, followed by protein synthesis, mitochondrial activity and total protein content (Wataha et al., 1991b, 1994b). These changes, which affect cell metabolism and protein production, generally occurred at lower Pd^{2+} concentrations in cultured macrophages than those that caused cell lysis (indicated by a loss of cellular membrane activity and measured by release of lactate dehydrogenase) (Wataha et al., 1995b). TC_{50} levels for palladium(II) chloride derived from tests of mitochondrial dehydrogenase activity (MTT assay) with mouse, hamster and human cell lines ranged from 134 to 360 µmol/litre (Wataha et al., 1991b, 1995b; Schmalz et al., 1997a).

DNA/RNA biosynthesis seems to be the most sensitive target. An EC_{50} value of palladium(II) chloride for inhibition of DNA synthesis *in vitro* with mouse fibroblasts was 300 µmol/litre (32 mg Pd^{2+}/litre) (Wataha et al., 1991b). This was consistent with the observed inhibition of DNA synthesis *in vivo* (Fisher et al., 1975; see also section 7.1). Mechanisms underlying the inhibition of DNA-mediated functions probably are dual, at least for palladium–organic complexes. They were found to inhibit transcription by altering both the DNA template and the enzymes involved (Mital et al., 1992).

Palladium applied in its metallic form (incubation of small test pieces) showed no cytotoxicity in mouse fibroblasts (Kawahara et al., 1968) or little cytotoxicity in human cell lines (Kawata et al., 1981; Niemi & Hensten-Pettersen, 1985), as evaluated microscopically. In a test system consisting of three-dimensional human fibroblast–keratinocyte co-cultures, palladium did not alter cell viability (measured by mitochondrial dehydrogenase activity, MTT assay) after 24 h of exposure. There was also no influence on prostaglandin E_2 release, but a 4-fold increase in interleukin-6 levels compared with untreated controls (Schmalz et al., 1997b, 1998).

Palladium-containing alloys tested varied in their *in vitro* cytotoxicity, depending mainly on microstructure and composition of the samples (Kawahara et al., 1968; Kawata et al., 1981; Niemi & Hensten-Pettersen, 1985; Ito et al., 1995; Warocquier-Clerout et al., 1995). Severe effects seem to be triggered by other components (e.g., copper). Exposure of a high-noble alloy (Au58, Ag25, Pd13, Zn4; weight per cent) to human fibroblast–keratinocyte co-cultures resulted in a 87–90% reduction of cell viability (MTT assay), but there was no change in prostaglandin E_2 or interleukin-6 levels (Schmalz et al., 1997b, 1998). Differences from controls in fibronectin arrangement and cell proliferation were observed with an alloy containing 78% palladium (Au2, Pt1, Pd78, Ag6, Sn2.5, In1.5, Ga9) in a human fibroblast culture (Grill et al., 1997).

7.8.2 Antineoplastic potential

The kind of cytotoxic potential mentioned above, the interference with DNA, is believed to predestine palladium compounds as antineoplastic agents. Therefore, many palladium–organic complexes were designed (mainly since 1986) and screened for cytostatic activity in order to obtain new anticancer drugs, having similar activity as *cis*-dichloro-2,6-diaminopyridine-platinum(II) (*cis*-platinum) but less adverse side-effects (e.g., Puthraya et al., 1985, 1986; Castan et al., 1990; Khan et al., 1991; Mansuri-Torshizi et al., 1991, 1992a,b; Mital et al., 1991; Teicher et al., 1991; Navarro-Ranninger et al., 1992, 1993; Higgins et al., 1993; Paul et al., 1993; Jain et al., 1994; Lee et al., 1994; Matilla et al., 1994; Curic et al., 1996). Concentrations required for growth inhibition of cancer cells should be less than those causing cytotoxic effects in normal cells. Generally, it is accepted that a palladium complex with useful antitumour potential should have an *in vitro* IC_{50} value (for growth inhibition) of about ≤ 10 µg/ml. IC_{50} values for potassium tetrachloropalladate(II) (as an example of an inorganic palladium complex salt) against human breast carcinoma cell lines were about 2 µg/ml (Navarro-Ranninger et al., 1992, 1993; Matilla et al., 1994).

To date, there are only few studies (Castan et al., 1990) that have tested or confirmed the antitumour activity of palladium complexes *in vivo*.

7.8.3 Enzyme inhibition

Many of the adverse effects of palladium compounds are mediated by binding to (see section 6.6) and inhibiting enzymes. Studies with a number of isolated enzymes from different animal systems and with distinct metabolic functions demonstrated that Pd^{2+} compounds had a high inhibition potential (Table 26). The strongest inhibition was found for creatine kinase (K_i value of 0.16 μmol/litre for palladium(II) chloride), an important enzyme of energy metabolism. In this case, inhibition was accompanied by a marked increase in electrophoretic mobility of the enzyme, possibly indicating conformational changes (Liu et al., 1979b). Another study suggests as a mechanism of inhibition the replacement of Fe^{2+} with Pd^{2+} at the active site of prolyl hydroxylase and the formation of strong complexes (Rapaka et al., 1976).

7.9 Other special studies

7.9.1 Nephrotoxicity

In addition to what has been noted in the previous sections in this chapter (i.e., change of kidney weight, glomerulonephritis), the nephrotoxic effects of three metal coordination compounds — namely, *cis*-platinum(II) (used for treatment of epithelial malignancies), palladium(II) chloride 2,6-diaminopyridine·H_2O (*cis*-palladium(II)) and *cis*-trichloro-2,6-diaminopyridine-rhodium(III)·$2H_2O$ (*cis*-rhodium(III)) (both exhibiting antimitogenic and antiviral properties) — were investigated by an *in situ* rat (Sprague-Dawley) kidney perfusion technique (Bikhazi et al., 1995). All three compounds showed comparable adverse effects on the major pumps and membrane proteins that are responsible for the transport of Na^+ and Ca^{2+} ions across kidney epithelial cells at concentrations of 0.56 mmol/litre. They inhibited both the Na^+-Ca^{2+}-antiporter and the Na^+-H^+-exchanger, with laxing effects on non-voltage-gated Ca^{2+} channels at the basolateral side. However, no significant effects were seen on the Na^+-K^+-ATPase and the Na^+-Ca^{2+} symporter.

Table 26. Inhibition of isolated enzymes by palladium compounds

Compound	Enzyme	Isolated from[a]	Inhibition[b]	Metabolic function[c]	Reference
PdCl$_2$	catalase	n. sp.	no		Spikes & Hodgson (1969)
	lysozyme	n. sp.	no		
	peroxidase	n. sp.	no		
	ribonuclease	n. sp.	no		
	α-chymotrypsin	n. sp.	yes: 500 µmol/litre (80% suppressed over 40 min)	degradation of proteins	
	trypsin	n. sp.	yes: 500 µmol/litre (at pH 4.2, but not at pH 8.9)	degradation of proteins	
PdCl$_2$	glutamic oxaloacetic transaminase (GOT)	white sucker (Catostomus commersoni)	yes: I_{20} = 50 mg (0.47 mmol)/litre	transamination	Christensen (1971/72)
	lactic dehydrogenase (LDH)	serum	yes: I_{20} = 35 mg (0.33 mmol)/litre	glycolysis	
PdCl$_2$	creatine kinase	rabbit muscle	yes: K_i = 0.16 µmol/litre; competitive; irreversible	energy metabolism (ATP regeneration)	Liu et al. (1979a,b)
		human serum	yes: to a lesser extent than above		
PdCl$_2$	alkaline phosphatase	calf intestine	yes: K_i = 0.6 µmol/litre; non-competitive	osteogenesis and fat absorption	Liu et al. (1979a,c)
PdCl$_2$	carbonic anhydrase	bovine erythrocytes	yes: K_i = 1.0 µmol/litre	acid–base electrolyte balance	Liu et al. (1979a)
PdCl$_2$	aldolase	rabbit muscle	yes: K_i = 4.0 µmol/litre; non-competitive	glycolysis	Liu et al. (1979a)
PdCl$_2$	succinate dehydrogenase	rat intestinal brush border	yes: K_i = 8.0 µmol/litre; non-competitive	TCA cycle	Liu et al. (1979a)

Table 26 (contd).

Compound	Enzyme	Isolated from[a]		Inhibition[b]	Metabolic function[c]	Reference
$PdCl_2$	prolyl hydroxylase	chick embryonic cartilage	yes:	K_i = 20 µmol/litre; competitive	collagen biosynthesis	Liu et al. (1979a)
$PdCl_2$	ribonuclease	bovine pancreas	yes:	I_{50} = 0.002 mmol/litre	nucleic acid metabolism	Christensen & Olson (1981)
$PdSO_4$	prolyl hydroxylase	(? chick cartilage)	yes:	K_i = 20 µmol/litre; competitive for the normal Fe^{2+} binding site	collagen biosynthesis	Rapaka et al. (1976)
$Pd(NO_3)_2$	aminopyrine demethylase	rat liver microsomes	yes:	0.2–0.3 mmol/litre (50% inhibition); non-competitive	drug metabolism	Holbrook et al. (1976)
K_2PdCl_6 ($Pd(IV)$)	ribonuclease	bovine pancreas	yes:	I_{50} = 0.05 mmol/litre	nucleic acid metabolism	Christensen & Olson (1981)
Na_2PdCl_6 $[Pd(NH_3)_4]Cl_2$ $(NH_4)_2PdCl_6$ K_2PdCl_6	cellobiohydrolase	fungus Trichoderma reesei	yes:	50 µmol/litre (84–85% inhibition)	hydrolysis of cellulose	Lassig et al. (1995)

[a] n. sp. = not specified.

[b] K_i = 50% inhibition of the enzyme activity (in µmol/litre); $I_{20\ (50)}$ = concentration causing 20% (50%) inhibition of enzyme activity.

[c] ATP = adenosine triphosphate; TCA = tricarboxylic acid.

7.9.2 Neurotoxicity

Symptoms such as ataxia, body tremors and tiptoe or splayed gait (observed in rats; Johnson Matthey, 1994a, 1995a; see also Table 21) after administration of high doses of palladium compounds may be indicative of some neurotoxic potential.

7.10 Toxicity of metabolites

There were no studies available on the toxicity of metabolites of palladium compounds.

7.11 Mechanism of toxicity/mode of action

The mode of action of palladium ions and of elemental palladium (as metal in alloys or as dispersed dust) in biological systems is not fully clear.

Similar to other transition metals, palladium ions follow a few basic principles in their mode of action. Due to their ability to form strong complexes with both inorganic and organic ligands, they have the potential not only to disturb cellular equilibria or replace other essential ions but also to interact with functional groups of macro-molecules, such as proteins (e.g., enzymes, membrane-bound transport proteins, etc.) or DNA/RNA, thereby disturbing a variety of cellular processes. Furthermore, different oxidation states may have different effects. Furthermore, the ability of Pd^{4+} ions to change their oxidation states to Pd^{2+} may contribute to harmful effects. The complete cascade of changes resulting in a particular toxic effect remains to be determined.

In some cases, palladium may act via formation of an aqua ion, $Pd(H_2O)_4^{2+}$ (Lassig et al., 1995). Such a complex is not formed with platinum. Generally, palladium ions have higher reaction rates (e.g., 10^3–10^5 times faster) than platinum ions (Kozlowski & Pettit, 1991).

No or little information is available on the mode of action of metallic palladium in its solid (alloys) or dispersed (dust particles) form in biological systems. It is assumed that, at least for processes in the oral cavity, ions can be formed from metallic palladium in dental restorations, leading to solution in the saliva and finally to distribution in the body. Fine particles of palladium may also be taken up by epithelial cells or lymphocytes and macrophages, etc., and be transported within the lymphatic system.

8. EFFECTS ON HUMANS

8.1 General population exposure

8.1.1 Effects due to exposure to palladium dust emitted from automobile catalytic converters

To date, no studies are available on the hazards of breathing fine particles containing palladium. As discussed in chapters 3 and 5, the emission of palladium dust (containing respirable fractions) from automobile catalysts may increase in future.

8.1.2 Effects after iatrogenic exposure

8.1.2.1 Dentistry

1) Case reports

There are several case reports on palladium sensitivity associated with exposure to palladium-containing dental restorations (Table 27). Symptoms observed included signs of contact dermatitis/stomatitis/mucositis and oral lichen planus. General symptoms, including swelling of the lips and cheeks, dizziness, asthma and chronic urticaria, have also been reported.

In some case reports, complaints disappeared after replacement with palladium-free (or metal-free) constructions. Nevertheless, the contribution of etiological factors other than palladium to patient symptoms cannot be ruled out. For example, patients showed frequently positive patch test reactions to other metals, mainly nickel salts.

Downey (1992) reported a case of acute intermittent porphyria after placement of dental prostheses composed of 76% palladium, 2% gold and 10% copper (probably by weight). The symptoms observed in a 45-year-old woman (skin rashes, diarrhoea, metallic taste, short-term memory loss, swelling of feet and arms, signs of chronic fatigue syndrome and others) disappeared following removal of the restorations. Despite the high palladium content, it was concluded that copper was the inducing agent, because copper is known to induce porphyria. The possible role of palladium was not discussed.

Table 27. Case reports on palladium sensitivity related to exposure to dental restorations

Subject, age (years)[a]	Exposure[b]	Symptoms	Positive skin patch tests with[c]	Additional reaction to	Other tests Type[d]	Result[e]	Reference
Woman (n = 1), 19	dental bridge (Pd: 5%)	relapsing painful swelling of the right cheek; pain in the oral cavity; generalized itching and dizziness; recovery after replacement	PdCl$_2$: 2% aq.	NiSO$_4$ CoCl$_2$	–	–	van Ketel & Niebber (1981)
Woman (n = 1), 68	dental prosthesis (Pd: n. sp.)	swelling of the lip; cutaneomucosal eczema; healing after replacement	PdCl$_2$: 1% pet.	none[f]	–	–	Castelain & Castelain (1987)
Woman (n = 1), n. sp.	numerous dental crowns (Pd: 79%)	severe erosive oral lichen planus	PdCl$_2$: 1% aq.	–	–	–	Downey (1989)
Person (sex not given) (n = 1), n. sp.	2 dental crowns (Pd: 79%)	painful lichen planus	PdCl$_2$: 1% aq.	–	–	–	Downey (1989)
Woman (n = 1), 28	dental bridge (Pd: 90%; Ni: 10%)	urticaria; swelling of the lower lips	PdCl$_2$: 2.5% aq.	NiSO$_4$	–	–	van Joost & Roesyanto-Mahadi (1990)
Woman (n = 1), 65	dental prosthesis (Au, Pt, Pd, Ru, Au; % n. sp.)	itching, tickling, dryness in the mouth	PdSO$_4$: 1% pet.	SnCl$_2$	–	–	Hackel et al. (1991)
Woman (n = 1), 54	dental prosthesis (Pd: 2%)	stomatitis; asthma; recovery after replacement	PdCl$_2$: 0.1–1% vas.	none	–	–	Kütting & Brehler (1994)
Woman (n = 1), 36	dental prosthesis (Pd)	stomatitis (biopsy from the test site of PdCl$_2$: eczematous and lichenoid changes)	PdCl$_2$: 1% pet.	(NH$_4$)$_2$PtCl$_4$	–	–	Koch & Baum (1996)

Table 27 (contd).

Subject, age (years)[a]	Exposure[b]	Symptoms	Positive skin patch tests with[c]	Additional reaction to	Other tests Type[d]	Result[e]	Reference
Woman (n = 1), n. sp.	alloy (Pd)	lichen ruber pemphigoides; recovery after replacement	PdCl$_2$; 1% pet.	none	–	–	Richter (1996)
Woman (n = 1), 74	dental bridge (Pd, Au, Ag, Cu)	oral lichen planus; disappearance after removal of bridge	PdCl$_2$; 1% (solvent n. sp.).	Sn, Ni, Fe, Sb salts	LST	(+)	Akiya et al. (1996)
Woman (n = 1), 24	dental alloy (Pd, Au, Mo, Hg)	atopic dermatitis; moderate improvement after removal of alloy	PdCl$_2$; 1% aq.	Au, Mo, Hg salts	–	–	Adachi et al. (1997)
Woman (n = 1), 47	dental prosthesis (ceramic Au–Pd alloy)	chronic urticaria-angio-oedema (on upper lip); no symptoms (for 18 months) after replacement	PdCl$_2$ (concentration and solvent n. sp.)	Ni, Co salts	–	–	Fernandez-Redondo et al. (1998)
Woman (n = 1), 26	dental amalgam, dental prostheses, dental stainless steel instruments	facial dermatitis, lichenoid buccal reactions	PdCl$_2$; 2% (solvent n. sp.)	Ni, Co, Cu salts, weak reaction to Au, Hg salts	–	–	Hay & Ormerod (1998)
Man (n = 1), 62	dental crowns and bridges (Pd–Ag–Au or Cr–Ni without Pt)	linear lichen planus from the cheek to the jaw (in the region of the mandibular nerve), unusual mouth sensation, itching; recovery after removal	PdCl$_2$ (concentration and solvent n. sp.)	Pt salt, but not to Ni and other metal salts	–	–	Mizoguchi et al. (1998)
Woman (n = 1), 70	dental crown (Pd)	1-year history of itchy erythematous eruptions over the mandibles, forehead and frontal scalp (contact dermatitis); improvement after replacement[g]	PdCl$_2$; 1% aq.	none	–	–	Katoh et al. (1999)

Table 27 (contd).

Subject, age (years)[a]	Exposure[b]	Symptoms	Positive skin patch tests with[c]	Additional reaction to	Other tests Type[d]	Result[e]	Reference
Man (n = 1), 39	dental alloy (20% Pd, 5% Ag, 17% Cu, 12% Au)	swelling of the local gingiva the next day after filling; exacerbation of bronchial asthma 3 months later (methacholin provocation test); improvement after removal	PdCl$_2$: 1% (solvent n. sp.) Pd metal	none of 20–30 allergens tested (including NiSO$_4$)			Yoshida et al. (1999)
Woman (n = 1), n. sp.	gold and amalgam fillings (Pd?[h])	itching and burning in the oral cavity, chronic fatigue syndrome	PdCl$_2$: 2% pet.	Au, Ni, Co salts	–	–	Stejskal et al. (1994)
Woman (n = 1), 39	dental gold constructions, amalgam fillings (Pd?[f])	mucosal problems, oral lichen (and a general history of allergy)	PdCl$_2$: 2% pet.	Au, Ni, Co salts	MELISA	+	Stejskal et al. (1994)

[a] n. sp. = not specified.
[b] Palladium content and other components, if specified (exact composition mostly not available), in parentheses; n. sp. = not specified.
[c] aq. = aqueous solution; pet. = petrolatum; vas. = vaseline.
[d] LST = lymphocyte stimulation test; MELISA = memory lymphocyte immunostimulation assay.
[e] – = not performed; (+) = unclear validity.
[f] none = none was positive.
[g] Recurrence of symptoms 1 day after fitting the original palladium prostheses; improvement after further replacement with palladium-free crowns.
[h] High-copper amalgams may contain also palladium (Eneström & Hultman, 1995).

112

Generally, it should be kept in mind that the possible detrimental effects of palladium in dental alloys may vary depending on overall composition and preceding preparation of the alloy, due to different corrosive behaviour (see chapter 5), different biocompatibility (see chapter 7) and (inter)actions with the other components.

2) Studies reporting the frequencies of palladium allergy

    In addition to case reports, there are numerous studies reporting the frequency of palladium allergy, as determined in special groups (with and without clinical symptoms) by means of patch test reactions to palladium(II) chloride. The frequency of sensitivity in apparently dentistry-related groups ranged from 3 to 36% (van Loon et al., 1986; Gailhofer & Ludvan, 1990; Koch & Bahmer, 1995, 1999; Tibbling et al., 1995; Marcusson, 1996; Richter & Geier, 1996; Schaffran et al., 1999), which was similar to the wide range found with eczema patients (see also Table 30 in section 8.1.4 below).

3) Special studies

    Several studies investigated the allergic potential of metallic palladium (pure or in mixture). As can be seen from Table 28, the epicutaneous or epimucosal tests performed resulted in only few positive reactions. Interestingly, many patients sensitive to palladium(II) chloride did not react to the applied metallic palladium. On the other hand, reactivity to metallic palladium was not necessarily connected with positive palladium(II) chloride patch test results.

    Several explanations for the weak reactivity towards metallic palladium in contrast to the reactivity to palladium(II) chloride have been suggested: there could be an additive effect of allergens (palladium and nickel) in tests with palladium(II) chloride containing traces of nickel contaminants (Todd & Burrows, 1992) or a mechanism of sensitization that requires the formation of either ions (Augthun et al., 1990) or specific complexes between ions and skin proteins (Santucci et al., 1995). Differences in the ionic charge (thus leading to different reactivity) may also play a role (De Fine Olivarius & Menné, 1992; Santucci et al., 1995). For example, $(PdCl_4)^{2-}$ seems to be less reactive than palladium(II) chloride (Santucci et al., 1995). The extent of release of palladium ions from foils onto skin is currently the most

Table 28. Tests of sensitivity to pure palladium and palladium alloy foils

Pd preparation (small foils[a])	Application[b] on skin	Application[b] in mouth	Test persons[c]	Reaction[d]	Remarks	Reference
Pure metal (purity: >98.7%)	–	×	Patient sensitive to $PdCl_2$, 2.5% aq.	0/3		van Loon et al. (1984)
Pure metal (purity: 99.99%)	–	×	Patient sensitive to $NiSO_4$	0/15	influence on number of T-lymphocytes; decrease of Langerhans cells (insignificant)	van Loon et al. (1988)
Pure metal (polished)	× ×(+ a.s.)	×(+ a.s.)	Patient sensitive to $PdCl_2$, 1% vas.	0/18 0/18 1/18 0/18		Augthun et al. (1990)
Pure metal (surface: 100% Pd)	×	–	Patient sensitive to $PdCl_2$, 1% pet.	0/19		De Fine Olivarius & Menné (1992)
	×	–	Dermatitis patients	3/470	none of these was positive to $PdCl_2$ or $NiSO_4$	Todd & Burrows (1992)
Pure metal (purity: 99.99%)	×	–	Patient sensitive to $PdCl_2$, 1% pet.	0/12		
Pure metal (purity: 99.99%)	×	–	Patient with diagnosed or suspected Ni allergy	1/103	no reaction to $PdCl_2$	Uter et al. (1995)
Pure metal (purity: 99.95%)	×	–	Patient sensitive to $PdCl_2$, 1% pet., and $NiSO_4$	0/87	3/87 reacted to $(PdCl_4)^{2-}$, aq.	Santucci et al. (1995)

Table 28 (contd).

Pd preparation (small foils[a])	Application[b]		Test persons[c]	Reaction[d]	Remarks	Reference
	on skin	in mouth				
Pure metal (purity: 99.95%)	×	–	Patient sensitive to $PdCl_2$, 1% pet.	1/1		Koch & Baum (1996)
Pure metal (purity: 99.95%)		×	Patient sensitive to $PdCl_2$, not sensitive to $NiSO_4$	1/1		Yoshida et al. (1999)
Pd–Ag alloy	–	×	Patient with "allergy"	1/2		Kratzenstein & Weber (1988)
Eight Pd-containing alloys	×	–	Patient with "suspected allergy"	4/141	reaction to one or more alloys	Mayer (1989)
Pd–Ag alloy	×	–	Patient with "possible allergy"	4/130		Schwickerath (1989)
Pd alloy (Pd76.5)	× (+/– a.s.)	× (+/– a.s.)	Patient sensitive to $PdCl_2$, 1% vas.	0/18; 0/18		Augthun et al. (1990)
Pd alloy (Pd79, Au2, Cu19)	×		Patient with severe local and systemic allergic reactions	1/1	reaction also to other alloys (Ni, Cr, Mo, Be)	Hansen & West (1997)

[a] Varying sizes: diameter of discs: 0.8–12 mm; sides of rectangles: 3–10 mm; thickness: 0.05–1 mm (if specified).
[b] a.s. = artificial saliva.
[c] aq. = aqueous solution; pet. = petrolatum; vas. = vaseline.
[d] Number of persons with clinical reaction/number of persons tested.

115

favoured explanation for the rates of sensitivity reported (Flint, 1998). However, the discordances are not yet fully understood.

No clear clinical effects on oral mucosa were found in 72 patients who wore partial dentures of palladium–copper–indium alloys (Pd73, Cu14, In5) for a period of up to 48 months (Augthun & Spiekermann, 1994), although this alloy showed a relatively low corrosion resistance (see chapter 5). Fourteen patients reported a metallic taste or "battery feeling" in the mouth. Another study noted slight or moderate reactions of the mucosa adjacent to prostheses consisting of palladium-type alloys in less than 20% of the 39 patients examined (Mjör & Christensen, 1993).

A study (see Table 30 in section 8.1.4 below) that found no visible clinical evidence of allergic stomatitis after contact with a pure palladium foil in patients allergic to nickel sulfate did find effects when the oral mucosa was examined immunohistologically. There were increases in the number of suppressor/cytotoxic T-lymphocytes in the connective tissue and a non-significant decrease in Langerhans cells in the epithelium adjacent to the palladium foil in a subgroup of 6 of 15 patients (van Loon et al., 1988).

The influence of a silver–palladium alloy (no composition reported) on humoral immunity was investigated in 22 persons with a normal medical history. Slight, but not significant, changes in serum IgA, IgG and IgM levels were seen 5–7 days (22 patients) and 20 days (5 patients) after placement of the new alloy restoration in five patients (Vitsentzos et al., 1988).

Recently, *in vitro* responses to palladium of lymphocytes from *in vivo* palladium-sensitized patients were measured by means of a modified lymphocyte transformation test, the so-called memory lymphocyte immunostimulation assay (MELISA). The study in Table 27 (Stejskal et al., 1994) indicated that palladium induced strong lymphocyte proliferation responses in patients with oral or systemic symptoms, but not in a similarly exposed unaffected person. However, the low specificity of this *in vitro* assay suggests that it is not useful for diagnosis of contact allergy to the metals gold, palladium and nickel, since a large number of false-positive results will be obtained (Cederbrant et al., 1997). Possibly, hypersensitivity is linked to certain genotypes, as suspected from studies with other metals (Stejskal et al., 1994; Eneström & Hultman, 1995).

8.1.2.2   Cancer therapy

Possible side-effects of treating various kinds of tumours, e.g., prostate cancer, with [103]Pd needles (in use since about 1987; see chapter 3) may refer to general symptoms of therapeutic (radioactive) irradiation and are not discussed in the context of this document. Altogether, there were no palladium-related complications reported that might preclude the use of [103]Pd needles in cancer radiotherapy (e.g., Sharkey et al., 1998; Finger et al., 1999).

According to Tomilets et al. (1980), palladium (and platinum) salts were shown to possess both histamine-releasing and histamine-binding properties. The latter effect might be one of the possible mechanisms of the antitumour effect of palladium as well as platinum salts, since histamine binding in tumour cells is suggested to suppress their proliferation.

8.1.2.3   Other therapeutic uses

Colloidal palladium hydroxide ($Pd(OH)_2$) was used for treating obesity (Kauffmann, 1913). For example, repeated subcutaneous injections of 2–10 mg palladium hydroxide caused weight losses of 4–19 kg in the treated persons ($n = 3$) in periods ranging from 10 days to 3 months. Even single doses of 50–100 mg of palladium hydroxide preparations were used. Side-effects noted were fever, euphoria, long-lasting discoloration and/or necrosis at the injection site. *In vitro*, a haemolytic effect of palladium hydroxide was found at a dilution of 1:25 000.

Oral dosages of about 65 mg palladium(II) chloride/day — given to (ineffectively) treat tuberculosis — produced no apparent adverse effects in tuberculosis patients (Meek et al., 1943).

Palladium(II) chloride (no concentrations reported) was topically applied as a germicide (Meek et al., 1943). There are no reports on possible side-effects.

Palladium(II) chloride (2%) was used without apparent toxic effect for cosmetic tattooing of the cornea of the eye (reviewed by Grant & Schuman, 1993).

Sensitization testing was not reported. For comparison, experimental (non-therapeutic) application of ammonium tetrachloropalladate(II) and allyl palladium chloride dimer (($C_3H_5PdCl)_2$) to intact inside forearm skin caused erythema and oedema as 24-h reactions, which was in good agreement with results from animal studies (Campbell et al., 1975; see also section 7.4.1).

8.1.3 Effects after exposure from other sources

Case reports on suspected palladium effects from non-iatrogenic or unclear exposures are summarized in Table 29. Skin disorders were the symptoms described. In all cases, palladium sensitivity, confirmed by means of patch tests, was accompanied by positive patch test reactions to nickel sulfate. In some cases, additional sensitization to other metals was found.

The results from skin reaction tests to pure palladium metal described with respect to dentistry (see section 8.1.2 and Table 28) are also of interest for non-iatrogenic exposures.

Individuals with a history of non-iatrogenic palladium exposure (e.g., via jewellery) may have contributed to the frequency of palladium sensitivity of some groups under study, described in Table 30 below.

8.1.4 Characteristics of palladium sensitivity

A number of studies (mostly those of serial patch testing) reported the frequency of palladium sensitivity in special groups consisting mainly of patients with dermatological or odontological problems. A few data were available on other groups (e.g., schoolchildren; more or less randomly selected persons). The basis of all studies, which are compiled in Table 30 in chronological order (by year of publication), is patch test reactions to palladium(II) chloride. Even if some complicating factors (e.g., irritant versus allergic patch test reactions) may be involved in patch testing of metal salts (Fowler, 1990), the results show interesting trends. The frequencies of palladium sensitivity ranged from

Table 29. Case reports on palladium sensitivity (identified by positive skin test results) related to non-iatrogenic or unclear exposure

Subject, age (years)	Exposure	Symptoms	Positive skin patch tests with	Additional reaction to	Other tests Type	Other tests Result[a]	Reference
Woman (n = 1), 35	ring (Pd: 90%, Ru: 10%; probably by weight); ring (Pt: 90%, Ir: 10%)	contact dermatitis of fingers	Pd metal (ring)[b]	NiSO$_4$; Pt metal; Pt 90/Ir10 alloy	–	–	Sheard (1955)
Man (n = 1), 29	firearms (Pd content: unknown)	erythematous vesicular lesions of the forearms	PdCl$_2$: 1% pet.[c]	NiSO$_4$	–	–	Guerra et al. (1988)
Man (n = 1), 27	metal weights (used for lifting exercises; Pd content unknown)	erythematous vesicular lesions of the neck, forearms and legs	PdCl$_2$: 1% pet.	NiSO$_4$	–	–	Guerra et al. (1988)
Woman (n = 1), 58	unknown	none	PdSO$_4$: 1% pet.	NiSO$_4$ CuSO$_4$	–	–	Hackel et al. (1991)
Women (n = 7), 19–45	unclear (metal jewellery?)	hand eczema	PdCl$_2$: 1% pet.	NiSO$_4$ (7) CoCl$_2$ (3) Hg metal (1)	–	–	Camarasa et al. (1989)
Women (n = 15), 19–68	unclear (metal jewellery?)	hand eczema	PdCl$_2$: 1% pet.	NiSO$_4$ (15) CoCl$_2$ (7) Hg metal (3)	LTT[d] (n = 4)	0	Camarasa & Serra-Baldrich (1990)
Girl (n = 1), 15	earrings (Pd) on both ear helices pierced 5 years earlier	persistent epitheloid granulomatous contact dermatitis on sites of piercing	PdCl$_2$: 1% pet.	CoCl$_2$ NiSO$_4$	e	e	Jappe et al. (1999)

Table 29 (contd).

a   – = not tested; 0 = negative.
b   Unclear if sensitization is elicited by palladium, ruthenium or iridium.
c   pet. = petrolatum.
d   LTT = lymphocyte transformation test.
e   Histological examinations of biopsies of the lesions and the positive patch test to $PdCl_2$ 1% pet. revealed dermal granulomas in both biopsies; additional tests (mycobacteriosis, sarcoidosis and foreign body reaction) negative.

Table 30. Summary of studies reporting the frequency of palladium sensitivity in special population groups (listed according to year of publication)[a]

Population group[b] (number)	Occurrence of positive reaction[c]	Test method (patch, unless otherwise specified)	Concomitant reaction to other metal salts	Remarks	Reference
Patients with a history of allergic reactions (17)	3/17 (17.6%)	PdCl₂: 2.5% aq.	Ni (3) Co (1)		van Loon et al. (1984)
Patients with a history of contact stomatitis (30) or dermatitis (16) or controls (17)	1/30 (3.3%) 5/16 (31.3%) 1/17	PdCl₂: 2.5% aq.	n.a.		van Loon et al. (1986)
Patients with Ni allergy (15)	6/15 (40%)	PdCl₂: 2.5% aq.	Ni (6)	strong reaction to Pd in 1/6	van Loon et al. (1988)
Patients with possible side-effects of dental materials (151; 119 f, 32 m)	2/151 (1.3%) f: 2	PdCl₂: 1% (vehicle n. sp.)	Ni (2) Co (2)	no oral mucosal affections[d]	Stenman & Bergman (1989)
Patients of a hospital for dermatology (486)	36/486 (7.4%)	PdCl₂: 1% vas. reading: day 2, 3	Ni (18) Ni + Co (16) Co (1)	monoallergic to Pd : 1	Augthun et al. (1990)
Contact dermatitis patients (5641; 3876 f; 1765 m); 1985–1989	f: 81/3876 (2.09%) m: 3/1765 (0.17%)	PdCl₂: 1% pet. (Ni: <2.5%)	Ni (97.5%) Co (40.7%) Cu (14.8%)	main sites of the mostly subacute type of eczema: hands, face, oral mucosa, forearms (51.8, 14.8, 9.9, 8.6%)	Gailhofer & Ludvan (1990)

Table 30 (contd).

Population group[b] (number)	Occurrence of positive reaction[c]	Test method (patch, unless otherwise specified)	Concomitant reaction to other metal salts	Remarks	Reference
Participants (95; 20 f, 75 m), randomly selected (73 dental students, 12 clinical staff, 7 patients)	0/95	$PdCl_2$: 1% aq. reading: day 2–3 (7)	n.a.	irritant reactions in 15/95	Namikoshi et al. (1990)
Patients with dermatitis (100)	10/100 (10%) 10/100 (10%)	$PdCl_2$: 1% aq. $[Pd(NH_3)_4](NO_3)_2$: 1% aq.	Ni (10) Ni (10)		Rebandel & Rudzki (1990)
Patients with contact dermatitis (hands, earlobes) without oral mucosal complaints (1521)	42/1521 (2.8%) f: 39, m: 3	$PdCl_2$: 1% pet. reading: day 2, 4	Ni (24) Ni + Co (12) Ni + Co + Hg (3)	monoallergic to Pd: 3	Camarasa et al. (1991)
Patients (no details) (1307; 815 f, 492 m)	32/1307 (2.5%)	$PdCl_2$: 1% pet. reading: day 2, 3, 7	Ni (29)		De Fine Olivarius & Menné (1992)
Patients of 21 dermatological practices in North Germany (580)	27/580 (4.6%)	$PdCl_2$: 1% vas.	n. sp.		Scheuer et al. (1992)
Patients with suspected contact dermatitis (536)	13/536 (2.4%)	$PdCl_2$: 1% pet. reading: day 3	Ni (13)		Todd & Burrows (1992)
Eczema patients with undefined previous Pd contact (1382)	115/1382 (8.3%)	$PdCl_2$: 1% pet. reading: day 2, 3	Ni (107)	monoallergic to Pd: 8; histological confirmation in 10 patients biopsied	Aberer et al. (1993)
Patients from several dermatology centres (3066)	180/3066 (5.9%)	$PdCl_2$: 1% pet.	Ni (131)	monoallergic to Pd: 49	J. Geier (unpublished results, 1993; cited in Uter et al., 1995)

Table 30 (contd).

Population group[b] (number)	Occurrence of positive reaction[c]	Test method (patch, unless otherwise specified)	Concomitant reaction to other metal salts	Remarks	Reference
Patients with Ni allergy (30)	13/30 (43%)	$PdCl_2$: 1% (vehicle n. sp.)	Ni (13)		Rudzki & Prystupa (1994)
Patients with suspected adverse reactions to prostheses (66; 44 f, 22 m)	4/66 (6%) f: 4	$PdCl_2$: 1.5% pet.	n. sp.		Vilaplana et al. (1994)
Patients with oral lichenoid lesions (11)	4/11 (36.4%)	$PdCl_2$: 1% pet. reading: day 1–17	Ni (2) Cr (2), Au (2), Zn (1)		Koch & Bahmer (1995)
Eczema patients from 14 Austrian centres (1992/1993) (11 544; f: 71.5%; m: 28.5%)	924/11 544 (8%) 1992 (f; m: 11.2%/2.8%); 1993 (f; m: 9.4%/1.6%)	$PdCl_2$: 1% vas. reading: day 2, 3	Ni (94.6%) Co (36.2%)	predominance of young patients	Kränke et al. (1995)
Patients of several German dermatology clinics (1993/1994: 2857/2583)	1993: 7.3% (f; m: 10%/3.2%); 1994: 7.5% (f; m: 11.3%/2.2%)	according to Schnuch et al. (1993)	n. sp.		Schnuch & Geier (1995)
Patients with suspected amalgam intoxication (15) and controls (46)	4/15 (27%) 7/46 (15%)	MELISA[e] MELISA[e]		difference between patients and controls not significant	Tibbling et al. (1995)
Patients with Ni allergy	190/309 (62%)	$PdCl_2$: 1% pet. reading: day 3 and/or later		at low Ni test concentrations, a higher frequency of concomitant Pd reactions	Uter et al. (1995)

Table 30 (contd).

Population group[b] (number)	Occurrence of positive reaction[c]	Test method (patch, unless otherwise specified)	Concomitant reaction to other metal salts	Remarks	Reference
Patients with allergic contact dermatitis; Italy: 1991–1992 (2300)	171/2300 (7.4%)	$PdCl_2$: 1% pet.	Ni (169) Co (2)	no history of occupational or dental metal exposure	Vincenzi et al. (1995)
Finnish schoolchildren, 14–18, orthodontologically treated, randomly selected (700: 417 f, 283 m)	48/700 (7%) f: **44/417 (11%)** m: 4/283 (1%)	$PdCl_2$: 2% pet. reading: day 3–5	Ni (45)	monoallergic to Pd: 3; piercing of ears (44 f, 1 m); orthodontic treatment (29 f, 1 m); eczema (41 f, 2 m)	Kanerva et al. (1996)
Patients claiming side-effects from dental alloy components (397: 326 f, 71 m)	30/397 (8%)	$PdCl_2$: 1% pet. reading: day 3, 4	Ni (7) Au (11) Au + Ni (10)	monoallergic to Pd: 2; delayed reactions to Pd (5/205); ranking of subjective symptoms[f]	Marcusson (1996)
Eczematous patients (112: 77 f, 35 m)	1/112 (<1%)	$PdCl_2$: 1% pet. reading: day 3	n. sp.		
Patients of 21 German dermatology clinics, 1991–1994 total (8000) m/d patients (756) stomatitis patients (402)	7.4% 9.4% 9.6%	$PdCl_2$: 1%	Ni (>80%)	frequency of clinical symptoms higher in patients monoallergic to Pd than in those allergic to Pd + Ni	Richter & Geier (1996)
Patients with allergic background (1000)	91/1000 (9.1%)	$PdCl_2$: 1% pet. reading: day 2, 4	Ni (66) Ni + Cr (21) Ni + Cr + Co (3)	monoallergic to Pd: 1	Santucci et al. (1996)

Table 30 (contd).

Population group[b] (number)	Occurrence of positive reaction[c]	Test method (patch, unless otherwise specified)	Concomitant reaction to other metal salts	Remarks	Reference
German schoolchildren and young adults with Ni sensitivity (recorded by 22 dermatology centres, 1990–1995) age 6–15 years age 16–30 years	1/9 (11.1%) 158/554 (28.5%)	$PdCl_2$: % n. sp. reading: day 3	Ni (1) Ni (158)		Brasch & Geier (1997)
Patients with symptoms of metal hypersensitivity (34; 30 f, 4 m)	18/34 (52.9%)	$PdCl_2$: 1% pet.	Au + Ni (19%) Au (13%)	monoallergic to Pd: 2; LTT and MELISA evaluated as not useful tests for diagnosis	Cederbrant et al. (1997)
Children and adolescents with suspected contact dermatitis (6–16 years; dermatology department, United Kingdom; 1991–1995)	4/9 (44.4%)	$PdCl_2$: 1% pet. reading: day 2 and 4	Ni (3)	acute dermatitis (at least 3); at some degree a history of pierced ears	Shah et al. (1997)
Patients with possible side-effects of dental restorations (194), 1991–1996; Germany	27/194 (13.9%)	$PdCl_2$: 1% pet. reading: day 3, 10, 17	Ni (13)	late appearing (day 10, 17) reactions: 7/27	Koch & Bahmer (1999)
Asymptomatic persons with and without gold dental restorations (136), USA	22/136 (16.2%)	$PdCl_2$: 1% pet. reading: day 2, 7	Ni (19) Au (10)	monoallergic to Pd: 2; no delayed reactions	Schaffran et al. (1999)

[a] Abbreviations used: aq. = aqueous solution; pet. = petrolatum; vas. = vaseline; f = female; m = male; m/d = patients suffering from oral mucosal complaints or those with suspected allergy to dental materials; n.a. = not applicable; n. sp. = not specified; LTT = lymphocyte transformation test; MELISA = memory lymphocyte immunostimulation assay.

[b] Bold letters designate a more or less randomly selected group (four entries).

125

Table 30 (contd).

c   Number positive/number tested and/or per cent positive.
d   Among 151 with possible oral or other side-effects of dental materials, there were 12 persons reporting oral problems; among those, no one was found to be
    sensitive to palladium, while two other persons (without oral symptoms) were found to be sensitive to palladium.
e   Palladium salt used not specified.
f   Frequencies in 101 patients: pronounced fatigue (62%), complaints in muscles and joints (52%); oral symptoms (20–40%; histologically verified oral lichen planus:
    4/101); however, specification on proportion of palladium-positive patients not given.

<1 to 44% (excluding persons with known nickel allergy), thus reflecting differences in subgroups tested, group size, test method (e.g., reading day: see Marcusson, 1996) or other variables. If subgroups of persons with known nickel allergy were included, maxima of over 60% were reached. Several recent and large studies from different countries (Austria, Finland, Germany, Italy) found very similar frequencies of 7–8% in patients with dermatological problems or in schoolchildren. When sex differences could be evaluated, a clear predominance in females was seen. As reported by Kränke et al. (1995), there was also a strong preponderance in younger patients, with a peak frequency between 21 and 30 years of age.

Generally, an association of palladium sensitivity to specific professions was difficult to establish from the statistical data provided, even if exposure may have been in some cases via occupation.

The study with the most participants (>10 000), involving the testing of about 25 allergens, demonstrated and confirmed that palladium belongs to the seven most frequently reacting sensitizers (second rank after nickel within metals; Fig. 1) (Kränke et al., 1995).

Noteworthy are the combined reactions to other metals, primarily to nickel (see Table 30 and Fig. 1). There are several explanations for the frequently observed palladium/nickel association, including nickel contamination of the patch test material and parallel multiple sensitizations or cross-reactions (due to similar chemical properties). The first possibility was ruled out (for most cases) by several authors (e.g., van Loon et al., 1984, 1986; van Joost & Roesyanto-Mahadi, 1990; Eedy et al., 1991). Support for the second thesis was given by the cases of solitary palladium or nickel allergy. Experiments with guinea-pigs (see section 7.4.3) demonstrated the potential for multiple sensitization and for cross-reaction (with palladium being the primary sensitizer). Studies on human T-lymphocytes found preliminary indications of cross-reactivity at the clonal level (Moulon et al., 1995; Pistoor et al., 1995). Santucci et al. (1996) suggested that concomitant reactions may be due to the immunoresponse towards antigenic complexes similarly formed by both metals. On the other hand, in another preliminary study, "palladium-specific" lymphocytes of patients with positive patch test reactions to palladium(II) chloride did not proliferate upon challenge with appropriate doses of nickel sulfate (Kulig et al., 1997; Schuppe et al., 1998).

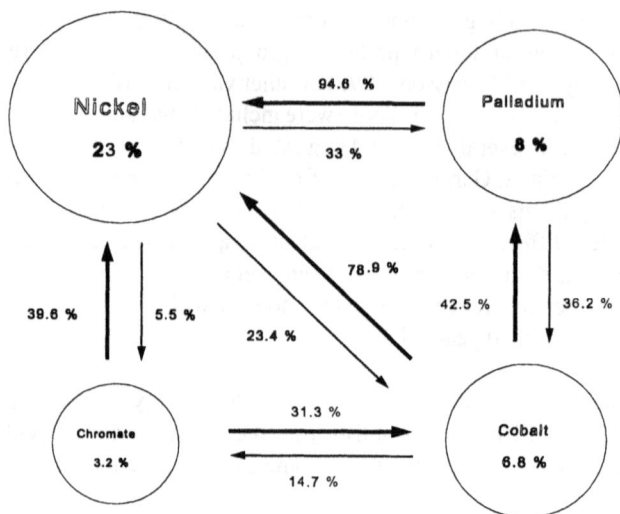

Fig. 1. Associated reactions of metals. Results from patch tests of >10 000 eczema patients (from Kränke & Aberer, 1996, with permission). Numbers in circles are the overall rates of sensitization; numbers on arrows are the rates of contemporaneous reactions, e.g., 94.6% of palladium-positive patients were also nickel-positive ($P < 0.0005$ for all associations).

Solitary (monoallergic) palladium reactions occurred at a low frequency (e.g., Augthun et al., 1990; Camarasa et al., 1991; De Fine Olivarius & Menné, 1992; Aberer et al., 1993; Uter et al., 1995; Kanerva et al., 1996; Marcusson, 1996; Santucci et al., 1996; Cederbrant et al., 1997; Schaffran et al., 1999). Case reports on solitary palladium allergy (e.g., Castelain & Castelain, 1987; Kütting & Brehler, 1994; Richter, 1996; Katoh et al., 1999; Yoshida et al., 1999) appear to be increasing in recent years.

It should be kept in mind that positive patch test reactions to palladium salts are not necessarily associated with oral symptoms or non-mucosal dermatitis. The clinical relevance remains unclear. Some authors point to the possibility of more systemic toxicity mediated by palladium-sensitized lymphocytes. Because these lymphocytes circulate freely in the blood and lymphoid system, they may possibly cause more distant effects (Stejskal et al., 1994; Marcusson, 1996).

Dental alloys and jewellery are a possible source of palladium sensitization in the general population.

8.2 Occupational exposure

8.2.1 Health effects due to metal (PGM) refinery processes

In a large-scale survey of South African platinum refinery workers (who are known to be exposed to palladium as well), positive skin prick test reactions to solutions of palladium halide salts dissolved in Coca's fluid were observed in 1 of 306 (Murdoch et al., 1986) or in 2 of 307 (Murdoch & Pepys, 1987) persons. The sensitization was confirmed by radioallergosorbent test and the monkey (*Macaca fascicularis*) passive cutaneous anaphylaxis test. The two palladium-positive workers also reacted to platinum salts at concentrations lower than those of the palladium salts. Another study using the monkey passive cutaneous anaphylaxis test also found a positive reaction to palladium salt (sodium tetrachloropalladate(II)) in 3 out of 22 platinum refinery workers with positive prick tests to platinum salts. In addition, heating experiments indicated the presence of heterogeneous antibodies (Biagini et al., 1985). As a mechanism for the observed concomitant reactions of palladium and platinum, a "limited" cross-reaction (rather than specific reactions or contamination of test material) was suggested (Murdoch & Pepys, 1987).

Roshchin et al. (1984) reported in a review article on the frequent occurrence of allergic diseases of the respiratory passages, dermatoses and affections of the eyes among Russian PGM production workers (further details not given).

8.2.2 Health effects due to use or processing of palladium-containing products

8.2.2.1 Dental technicians

According to an older study (Menck & Henderson, 1976), dental laboratory technicians were classified as an occupational group with a significantly increased rate of lung cancer (standardized mortality ratio = 4.0, $P < 0.01$). Similarly, a high risk for pneumoconiosis due to dust exposure was reported (Augthun et al., 1991, and references therein). Although some investigators did find palladium in respirable dust particles of dental laboratories (see section 5.3), the contribution of palladium (within a series of other substances generated during dental working processes) to the above-mentioned health hazards is not clear.

8.2.2.2   Automobile industry workers

Only one study focusing on workers processing or handling automobile catalysts is available. Prick tests with palladium(II) chloride were performed in 1990–1991 in workers of a German plant manufacturing automobile catalysts (Merget, 1991). Four out of 130 workers showed positive reactions to palladium(II) chloride (1 mmol/litre, ~177 mg/litre). They also reacted to hexachloroplatinic acid (H_2PtCl_6). (Corresponding workplace exposure data for the PGMs are not available.)

8.2.2.3   Others

Some cases of allergic dermatitis due to palladium (and possibly other metals) are documented for two chemists and a metal worker (Table 31).

A case has been reported of occupational rhinoconjunctivitis and asthma due to an isolated sensitization to palladium in a worker of the electronics industry (Daenen et al., 1997, 1999). About 30 min after a brief exposure to the fumes of an electrolysis bath containing palladium, used to coat electronic parts, a previously healthy, non-smoking, non-atopic, 26-year-old male developed (transient) symptoms of conjunctivitis, rhinitis, chest tightness and dyspnoea. Pulmonary function tests (peak flow records) confirmed the existence of asthma. Usual causes of allergy were not found in this worker. He was also exposed to other metal baths (nickel, tin, lead, gold), but not to platinum, whose salts are well known to cause asthma. Skin prick tests with tetraammine palladium(II) chloride (0.001%) as well as a bronchial provocation test to aerolized tetraammine palladium(II) chloride (0.0001–0.001%, several times for 5–180 s) were positive. The latter gave an early reaction (forced expiratory volume in 1 s [FEV_1] – 35%) and no late change in histamine PC_{20} (1.2 mg/ml; provocation concentration that causes a 20% fall in FEV_1). Exposure of a control asthmatic subject to tetraammine palladium(II) chloride gave no reaction. Skin prick tests carried out with solutions of sodium hexachloroplatinate(IV) (Na_2PtCl_6), ammonium tetrachloroplatinate(II) ((NH_4)$_2PtCl_4$) and palladium(II) chloride were negative for the platinum salts (up to 1%) and possibly

Table 31. Case reports on palladium sensitivity[a] associated with occupational exposure

Occupation (number, sex, age in years)	Exposure to	Symptoms	Positive skin patch tests with	Concomitant reaction to other metal salts	Reference
Chemist, (1, male, 25)	various precious metals (including Pd)	hand and forearm dermatitis	Na_2PdCl_2 (probably Na_2PdCl_4): 0.1, 1% (vehicle not specified)	$NiSO_4$	Munro-Ashman et al. (1969)
Chemist, (1, female, 37)	various metals (Ni, Cr, Co, Pd) for experimental electrolytic coating; earlier: episode of earring dermatitis	hand dermatitis	$PdCl_2$: 0.5–1% aqueous solution [Pd(NH$_3$)$_4$](NO$_3$)$_2$: 1% aqueous solution	Ni, Cr, Co	Rebandel & Rudzki (1990)
Metal worker (1, male, 29)	unclear (not aware of handling Pd-containing objects)	hand dermatitis	$PdCl_2$: 1% petrolatum	$CoCl_2$ $CoSO_4$	Hackel et al. (1991)

[a] Identified by means of patch tests; for additional case reports based on skin prick tests (indicative of respiratory sensitization), see text in sections 8.2.2.2 and 8.2.2.3.

positive for palladium(II) chloride (0.1%). In a second series of skin prick tests (performed more than 1 year later), the positive response to tetraammine palladium(II) chloride was confirmed. Tests with other salts (nickel chloride, cobalt chloride, ammonium hexachlororhodanate ($(NH_4)_3RhCl_6$), platinum salts, ammonium tetrachloropalladate(II) and ammonium hexachloropalladate(IV)) were negative, surprisingly including the two additional palladium compounds (Daenen et al., 1999).

8.3 Subpopulations at special risk

People with known metal (especially nickel) allergy may have an increased risk of palladium allergy (see Fig. 1).

8.4 Carcinogenicity and other effects

There are no data on carcinogenicity, reproductive toxicity or other effects in humans.

9. EFFECTS ON OTHER ORGANISMS IN THE LABORATORY AND FIELD

9.1 Laboratory experiments

9.1.1 Microorganisms

Several palladium compounds have been found to have antiviral, antibacterial and/or fungicidal properties.

Chloropalladosamine, ammonium tetrachloropalladate(II) (Graham & Williams, 1979) and $PdCl_2(2,6$-diaminopyridine)$\cdot H_2O$ (Tayim et al., 1974) were reported to be toxic to viruses. Bactericidal activity was seen with allyl palladium chloride dimer (($\pi C_3H_5PdCl)_2$) (Graham & Williams, 1979), several palladium(II) mixed ligand complexes (Khan et al., 1991) and complexes with orotic acids (Hueso-Urena et al., 1991). "Soluble Pd (anion n. sp.) salt" (White & Munns, 1951), palladium(II) nitrate (Somers, 1959) and allyl palladium chloride dimer (Graham & Williams, 1979) produced fungitoxicity. Concentrations required for the antimicrobial effects were mostly in the range of 12.5–2000 mg/litre. Palladium(II) chloride (at 10^{-5}–10^{-3} mol/litre) reduced the conversion yield of L-malate from glucose by *Schizophyllum commune* (Tachibana et al., 1972). Formerly, palladium(II) chloride was used as a germicide (Meek et al., 1943). Some palladium(II) complexes showed broad-spectrum antimicrobial activity against some human pathogens at (4.6–9.1) × 10^{-4} mol/litre (Khan et al., 1991).

Similarly, effects of several palladium compounds on microorganisms in the environment can be expected. However, few results from standard microbial toxicity tests under environmentally relevant conditions are currently available.

Recently, the inhibitory effect of tetraammine palladium hydrogen carbonate on the respiration of activated sewage sludge has been assessed according to OECD Guideline No. 209. The test resulted in a 3-h EC_{50} of 35 mg/litre (12.25 mg palladium/litre) (Johnson Matthey, 1995e).

9.1.2 Aquatic organisms

9.1.2.1 Plants

An algal cell multiplication inhibition test has been performed according to OECD Guideline No. 201, investigating the effect of tetraammine palladium hydrogen carbonate on the growth of *Scenedesmus subspicatus* over a 72-h period (Johnson Matthey, 1997e). It resulted in a 72-h EC_{50} value of 0.066 mg/litre (0.02 mg palladium/ litre) (reduction of biomass) and a 24-h EC_{50} value of 0.078 mg/litre (0.03 mg palladium/litre) (reduction of growth rate), based on nominal concentrations. The no-observed-effect concentration (NOEC) at 72 h was 0.04 mg/litre (0.014 mg palladium/litre). It was not possible to calculate EC_{50} values, etc., based on measured concentrations, which were low and variable, ranging between less than the limit of quantitation and 59% of nominal concentrations.

A number of PGM complexes have been tested for toxicity to the water hyacinth (*Eichhornia crassipes* (MART.) Solms) (Farago & Parsons, 1985a,b, 1994). Both palladium compounds tested, potassium tetrachloropalladate(II) and chloropalladosamine, turned out to be very toxic. If present in nutrient solution at 2.5 µg palladium/ml for 2 weeks (renewal after 1 week), they caused chlorosis and a drop in yield. At applied concentrations of 10 µg palladium/ml, the symptoms became more marked, with some necrosis and stunted dark roots. Visual appraisal of the degree of necrosis in the roots and tops of the plants ($n = 4$ per group) exposed to the various complexes at 10 µg metal/ml resulted in the following relative order of toxicity: Pt(II), Pd(II) > Ru(III) ≈ Ru(II) ≈ Ir (III) > Pt(IV) ≈ Os(IV) >> Rh(III), and this order correlated (except for iridium) with metal uptake and translocation (see also section 4.1).

9.1.2.2 Invertebrates

The acute toxicities of palladium and an additional 31 metal ions to *Tubifex tubifex* (Müller), a freshwater tubificid worm (Annelida, Oligochaeta), which is an important link in aquatic food-chains, have been tested according to standard methods (Khangarot, 1991). The 96-h EC_{50} values ranged from 0.0067 (osmium tetroxide) to 812.8 (potassium chloride) mg/litre. Palladium, with an EC_{50} value of 0.092 mg/litre, was one of the most toxic ions (sixth place after osmium, silver, lead, mercury and platinum). As seen in Table 32, palladium was only slightly less toxic than platinum.

Table 32. Acute toxicities of palladium and platinum ions to
Tubifex tubifex (Müller)[a,b]

Exposure duration (h)	EC$_{50}$ and 95% confidence limits (mg metal/litre)	
	Pd^{2+} (PdCl$_2$)	Pt^{2+} (PtCl$_2$)
24	0.237 (0.183–0.316)	0.095 (0.086–0.163)
48	0.142 (0.107–0.184)	0.086 (0.073–0.092)
96	0.092 (0.033–0.052)	0.061 (0.050-0.079)

[a] From Khangarot (1991).
[b] Tubificid worms were collected from natural sources and acclimatized for 7 days; n = 10 per concentration; three replicates per concentration; renewing of test water every 24 h; further test conditions according to APHA et al. (1981).

Acute toxicity to *Daphnia magna* has been assessed for tetra-ammine palladium hydrogen carbonate according to OECD Guideline No. 202. The 48-h EC$_{50}$ (immobilization) based on nominal test concentrations was 0.22 mg/litre (0.08 mg palladium/litre) (with 95% confidence limits of 0.20–0.25 mg/litre) (0.01–0.09 mg palladium/litre). The NOEC was 0.10 mg/litre (0.05 mg palladium/litre). Due to a considerable variability in the corresponding measured test concentrations, an EC$_{50}$ value based on the time-weighted mean measured concentrations was also calculated. This 48-h EC$_{50}$ value was 0.13 mg/litre (corresponding to 0.05 mg palladium/litre), with 95% confidence limits of 0.11–0.14 mg/litre (0.04–0.05 mg palladium/litre). The NOEC was 0.06 mg/litre (corresponding to 0.02 mg palladium/litre) (Johnson Matthey, 1997f).

9.1.2.3   Vertebrates

The minimum 24-h lethal concentration of palladium(II) chloride to cyprinodont freshwater fish medaka (*Oryzias latipes*; n = 3 per group) has been reported to be 0.04 mmol/litre (7 mg/litre [4.2 mg palladium/litre]). Compared with platinum (hexachloroplatinic acid: 0.08 mmol/litre), palladium was more toxic (Doudoroff & Katz, 1953).

Toxicity data from the exposure of rainbow trout (*Oncorhynchus mykiss*) to tetraammine palladium hydrogen carbonate at concentrations of 0.01, 0.1, 0.18, 0.32, 0.56, 1, 10 and 100 mg/litre are available from studies of Johnson Matthey (1997g). The method followed that described in OECD Guideline No. 203. There were 100% (3/3) mortalities in the 1, 10 and 100 mg/litre test groups. The 96-h LC_{50} was determined (under semistatic test conditions) to be 0.53 mg/litre (corresponding to 0.19 mg palladium/litre), with 95% confidence limits of 0.44–0.64 mg/litre (0.15–0.22 mg palladium/litre). The NOEC was 0.32 mg/litre (corresponding to 0.11 mg palladium/litre). Sublethal effects such as increased pigmentation and loss of equilibrium have been observed 24–96 h after exposure of fish ($n = 4$–10) to 0.56 mg/litre (corresponding to 0.20 mg palladium/litre). All effect concentrations were given as nominal concentrations, even if measured test concentrations at 96 h varied from 75 to 90% of nominal.

9.1.3   Terrestrial organisms

9.1.3.1   Plants

Effects of various concentrations of palladium(II) chloride on Kentucky bluegrass (*Poa pratensis* L.) grown in a nutrient medium were determined over 4 weeks (Benedict, 1970; Sarwar et al., 1970). Whereas small quantities of palladium(II) chloride stimulated growth, high concentrations caused the plants to die 1 week (115 mg/tray or 100 mg/litre [60 mg palladium/litre]) or 2 days (575 mg/tray or 500 mg/litre [300 mg palladium/litre]) after addition of palladium(II) chloride. At 3 mg/litre (1.8 mg palladium/litre) and above, inhibition of transpiration was observed; at 10 mg/litre (6 mg palladium/litre), histological changes (including aberrant stomatal histogenesis, inhibition of nodal meristem development, changes in chloroplast structure and hypertrophy of mesophyll cells, nuclei and nucleoli) became apparent. However, it was not clear (Sarwar et al., 1970) if the phytotoxicity was due to the palladium ion or to a non-specific ionic effect.

Detrimental effects of palladium(II) chloride were also found by an early study (Brenchley, 1934) testing several crop plants (barley, wheat, oats, peas, beans). Dose-dependent growth retardation and stunting of the roots were temporary at low, and more persistent at higher, concentrations of palladium(II) chloride added to the nutrient solution. The tolerance of palladium varied with species, the most sensitive being oats (affected at about 0.22 mg palladium(II) chloride/ litre [0.132 mg palladium/litre]).

9.1.3.2   Invertebrates

No data are available on the effects of palladium on terrestrial invertebrates.

9.1.3.3   Vertebrates

No data are available on the effects of palladium on terrestrial vertebrates.

9.2   Field observations

There are no data available on the effects of palladium on organisms in the field.

10. EVALUATION OF HUMAN HEALTH RISKS AND EFFECTS ON THE ENVIRONMENT

10.1 Evaluation of human health risks

10.1.1 Exposure levels

Generally, the very few data available do not allow a representative picture to be provided. Nevertheless, some trends may become apparent.

10.1.1.1 General population exposure

The intake of palladium from food or drinking-water is low. For drinking-water, a maximum daily intake of 0.03 µg palladium/person per day has been calculated (assuming a consumption of 2 litres/day). According to a United Kingdom survey, the total daily dietary intake of palladium has been estimated to be up to 2 µg/person per day. There may be a higher intake in some population groups consuming diets with high palladium levels (e.g., some types of mussels).

With dental alloys, additional oral exposure has been documented. Palladium in saliva may reach concentrations higher than 10 µg/litre and thus contribute considerably to the total palladium intake (<1.5–15 µg/person per day, assuming a production of 1.5 litres of saliva per day) in subjects having dental alloys. Substantial individual variation exists. The palladium-containing dental alloys exhibit a complex release behaviour that cannot be predicted from their nominal composition.

There are few data available concerning the actual palladium concentration in ambient air (1 and 3 pg/m^3, 13 and 57 ng/m^3). The two latter values are unexpectedly high. The authors of the study giving the latter two values remarked that this was a limited study with few samples. Furthermore, the analytical method for palladium was not specified. However, if an analogy to platinum (see IPCS, 1991) is assumed, ambient air levels may be ≤100 pg/m^3 in urban areas when palladium catalysts in automobiles are used. Therefore, the palladium uptake via inhalation from this source is also expected to be low (about

2.2 ng/person per day, assuming an average daily inhalation volume of 22 m^3). On the other hand, if the worst case (57 ng/m^3) is considered, the calculated daily intake would be 1254 ng/person per day.

Roadside dust levels ranged from 1 to about 300 µg palladium/kg. Estimates of human intake are not available from this source, but it is expected to make only a minor contribution.

Skin or mucosal contact with palladium-containing jewellery and dental alloys appears to be an important route of exposure. Internal exposures have not been accurately quantified to date. Subgroups with enhanced contact to palladium (and other metals), such as individuals with pierced ears or other body parts, may be at special risk to develop sensitivity to palladium.

10.1.1.2 Occupational exposure

If there are no protective measures (e.g., masks, gloves), dental technicians may be exposed to maximum levels of 15 µg palladium/m^3 air, leading mainly to inhalative, but probably also to skin, exposures.

Exposures to palladium or palladium compounds (in the form of dusts or solutions) may also occur in workers of the palladium mining, smelting, refining or recycling industries, in the chemical industry, particularly in catalyst manufacture, in the electronics industry or in jewellery/optical instruments fabrication.

From the limited available data, it appears that occupational exposure levels reported as palladium in industrial operations in which palladium compounds are handled would be similar to those levels in the dental laboratories mentioned above. It should be noted that the main exposure in chemical operations is to palladium salts.

There is one study reporting measurable levels of palladium (mean 1.1, maximum 7.4 µg/litre) in urine of refinery workers exposed to up to 0.36 µg palladium/m^3 air. In another study, urine levels of 8.4–1236 ng/litre were reported in dental technicians. Background levels for human urine are usually <0.1 µg/litre.

10.1.2   Fate in the body

Palladium ions can be taken up by the skin or by oral and inhalative routes. Although absorption and retention are poor, there may be a risk for sensitized or sensitive persons. Under certain conditions, palladium ions (and possibly microparticles) appear to be released from metallic palladium (e.g., in dental alloys). There are also indications that very finely dispersed elemental palladium particles become bioavailable when dissolved in biological media. However, precise quantitative data are not available. Principally, absorbed palladium can be found in almost all organs, tissues or body fluids, with maxima in kidney, liver and spleen (of experimental animals). The biological half-life of palladium in rats has been estimated to be 12 days.

10.1.3   Health hazards

Generally, it is hypothesized that the effects of palladium metal are mediated via the presence or release of palladium ions.

Palladium and its compounds are of very low to moderate acute toxicity if swallowed (depending mainly on their solubility).

Several palladium salts may cause severe primary skin and eye irritations.

An important target is the immune system. Palladium ions have been shown to be potent skin sensitizers. Epidemiological studies demonstrated that palladium ions are among the most frequent sensitizers within metals (second rank after nickel). Persons with known nickel allergy often display palladium allergy. Almost all persons with palladium allergy are also sensitive to nickel. There are also some indications of a potential for respiratory sensitization.

Dental alloys and jewellery are a possible source of palladium sensitization in the general population.

Palladium ions are capable of eliciting *in vitro* a series of cytotoxic effects. Little *in vitro* cytotoxicity has been seen after exposure to metallic palladium.

No mutagenic activity of palladium salts has been found in bacterial test systems. However, one compound (tetraammine palladium hydrogen carbonate) produced a weak clastogenic *in vitro* response with human lymphocytes. Interactions with DNA have been observed *in vitro*. An inhibition of DNA synthesis has been demonstrated both *in vitro* and *in vivo*.

The carcinogenic risk from oral exposure to palladium salts or through palladium-containing metallic implants is unclear. Only two animal experiments, with considerable limitations, have been performed, one reporting an increased occurrence of malignant tumours in mice exposed to palladium via drinking-water for a lifetime, and the other reporting tumours at the implantation site of a palladium-containing alloy in rats. No carcinogenicity studies after inhalation exposure are available.

Some general toxic effects, such as delayed body weight gain, changes in parameters of clinical chemistry or pathological changes in inner organs of experimental animals, have been recorded not only following oral doses of palladium salts, but also after oral treatment with elemental palladium.

10.1.4 Dose–response relationships

A major source of concern is the sensitization risk. The available data from animal and human findings do not allow a calculation of a NOAEL for sensitization in humans. Principally, however, it must be taken into consideration that very low doses are sufficient to cause allergic reactions in susceptible individuals. Persons with known nickel allergy may be especially susceptible.

A 28-day study in rats focusing on histopathological end-points (treatment-related abnormalities in liver, spleen, kidney and gastric epithelium of rats after oral dosing) reported a NOAEL of 1.5 mg tetraammine palladium hydrogen carbonate/kg body weight per day (corresponding to 0.54 mg palladium/kg body weight per day). However, absolute organ weight changes occurred at this dose level.

The only available 5-month inhalation study reported effects at both exposure levels of chloropalladosamine used (5.4 and 18 mg/m^3, corresponding to 2.6 and 8.7 mg palladium/m^3, respectively) with respect to clinical parameters recorded in rats. The study did not include histopathological examinations. Inhalation studies with elemental palladium are lacking, although palladium-containing particles are generated in dentistry and other workplaces or in the ambient air from automobile catalysts.

A lifetime study with mice given palladium(II) chloride in drinking-water at a dose of about 1.2 mg palladium/kg body weight per day found a significantly higher incidence of amyloidosis in several inner organs of males and females (besides a questionable carcinogenic effect) and suppressed growth in males, but not in females.

10.1.5 Health-based guidance value

Due to the lack of suitable data, it is not possible to make a quantitative risk assessment. Therefore, no health-based guidance value can be derived.

10.2 Evaluation of effects on the environment

10.2.1 Exposure levels

Palladium occurs naturally in the Earth's crust at widely varying concentrations (a median value of <1 µg/kg in the upper continental crust is estimated). Higher levels have been found in other environmental matrices, such as dust along roads and sewage sludges. For example, concentrations of up to 280 µg palladium/kg have been recorded in urban road dust, values being strongly correlated to the density of traffic. Generally, it can be expected that environmental palladium concentrations will increase with increasing use of palladium catalysts in automobiles.

Municipal sewage sludges from jewellery industry areas contained palladium concentrations of up to 4.7 mg/kg dry weight. A maximum concentration of 4 mg/kg dry weight has also been recorded in a highly polluted river sediment. Data on palladium emissions into wastewaters and surface waters or into air in the vicinity of industry are lacking. Similarly, palladium concentrations inside and outside of waste dumps and in the vicinity of incinerators have not been monitored.

Sporadic data are available for biotic media, suggesting the bioavailability of palladium. For example, plants grown on palladium-contaminated soil near a highway contained palladium at concentrations of up to about 2 μg/kg dry weight. Palladium concentrations in single samples of a freshwater plant (water hyacinth) and a feral pigeon were in the range of 100–800 μg/kg dry weight. Marine invertebrates (spot prawn) showed maximum palladium concentrations of 6 mg/kg dry weight.

10.2.2 Persistence, fate and transport

Elemental palladium — like other noble metals — tends to persist in the upper soil layer. Under appropriate conditions, it may become bioavailable. Small airborne palladium-containing particles may be distributed to other environmental media.

Palladium in its ionic form is assumed to be mobile in soil and water. Transfer from the water column to sediment may occur through exchange, complexation or precipitation reactions.

Methylation reactions have been observed with palladium salts, but it is uncertain whether these or other transformations occur in the environment.

There is limited information available to indicate that palladium may be accumulated in biota.

The fate of palladium in aquatic and terrestrial food-chains is not clear.

10.2.3 Toxicity and dose–effect/response relationships

The few studies available reported toxicity values based upon nominal exposure concentrations. Where exposure concentrations were measured, there were discrepancies with the nominal concentrations. Palladium salts have been found to be very toxic to aquatic plants, invertebrates and vertebrates (fish), showing LC_{50}/EC_{50} values below 1 mg palladium/litre. The most sensitive group has been green algae, with EC_{50} values (for inhibition of cell multiplication) of <0.1 mg palladium/litre. Detrimental effects (growth retardation, stunting) were also seen with terrestrial plants exposed to concentrations as low as

0.22 mg palladium(II) chloride/litre (0.13 mg palladium/litre). A single test with sewage sludge bacteria resulted in lower, but also considerable, toxicity (EC_{50} 12.3 mg palladium/litre for respiration inhibition).

No information is available for other indicator species.

There is also a lack of data on short-term or long-term ecotoxicity.

The effects of elemental palladium in a finely dispersed form under environmental conditions have not been investigated.

10.2.4 Guidance value

Due to lack of suitable data, it is not possible to make a risk evaluation, and therefore no guidance value can be recommended.

11. CONCLUSIONS AND RECOMMENDATIONS FOR PROTECTION OF HUMAN HEALTH AND THE ENVIRONMENT

11.1 Dental health

Since palladium-containing dental alloys have been identified as a possible source of sensitization, protection of the public from related adverse effects may be achieved either by limiting the use of certain alloys or by using alloys with minimal release of palladium.

Dentists should be informed of the composition of alloys and of the possible sensitization effects of palladium.

Patients should be informed about the composition of dental alloys. Patients who have an allergy to nickel should be informed that the use of palladium-containing dental materials may cause palladium allergy, but that this risk appears to be low, and the potential for this reaction is likely to vary with the release of palladium ions from the material. In general, in patients who are sensitive to palladium, palladium-containing materials should not be used, although palladium has been used without allergic effects in some of these individuals.

11.2 Occupational health

Occupational exposure to palladium metal, alloys and compounds occurs in conjunction with exposure to other PGM alloys and compounds. Guidelines on the medical surveillance of workers exposed to platinum salts have been published (e.g., HSE, 1983), but no guideline specific to palladium exposure has been published.

In general, people with known palladium allergy should not work with palladium compounds. People with known nickel allergy should be advised that working with palladium salts may cause allergic reactions.

Pre-employment screening should include a questionnaire for skin disease specifically for allergy to metals (nickel, cobalt and palladium). Routine patch tests with nickel and palladium compounds are not recommended. Patch tests should be performed only to determine the• cause of occupational dermatitis.

During employment, regular checks should be made for skin and respiratory health by questionnaires and examinations.

Personal protective equipment should be used to prevent skin contact with palladium compounds.

11.3 Analysis

Due to its widespread use as a chemical modifier for the analysis of trace elements in biological matrices by GF-AAS, care must be taken to avoid contamination when measuring palladium by the GF-AAS technique.

11.4 Environment

Palladium ions are considered to be highly toxic to aquatic organisms. However, due to palladium's high economic value, emissions of palladium from point sources are currently minimal. Increased use of catalytic converters may increase palladium emissions from diffuse sources. These emissions should be controlled to be as low as possible.

12. FURTHER RESEARCH

- Characterization of palladium particles (including their size distribution and soluble fraction) occurring at workplaces and in the general environment, including emissions from automobile catalytic converters.

- Providing certified reference materials and performing inter-laboratory comparisons and introduction of quality assurance programmes.

- Measurements and documentation of occupational exposures for a variety of workplaces. Appropriate (longitudinal) epidemiological studies of the incidence of allergic (and other) diseases in palladium-exposed workers.

- Further studies on release and bioavailability of palladium from dental alloys and jewellery.

- Further research on the role of palladium in dermal/mucosal sensitization, respiratory sensitization and other allergic reactions, including cross-reactivity at the clonal level.

- Adequate comparative studies on long-term toxicity (including carcinogenicity) of palladium administered in its different valencies via relevant routes (oral, inhalative, dermal/mucosal exposure). Special studies addressing the potential for neurotoxicity and immunotoxicity, reproductive toxicity and genotoxicity.

- Kinetic studies (including validation of body fluids as suitable indicators of palladium exposure).

- Studies on the long-term toxicity of palladium compounds and finely dispersed elemental palladium on aquatic and terrestrial test organisms (including sediment- and soil-inhabiting species).

13. PREVIOUS EVALUATIONS BY INTERNATIONAL BODIES

No international evaluations were available.

REFERENCES

Abbasi SA (1987) Analysis of sub-microgram levels of palladium(II) in environmental samples by selective extraction and spectrophotometric determination with N-p-methoxyphenyl-2-furylacrylohydroxamic acid and 5-(diethylamino)-2-(2-pyridylazo) phenol. Anal Lett, **20**: 1013–1027.

Aberer W, Holub H, Strohal R, & Slavicek R (1993) Palladium in dental alloys — the dermatologists' responsibility to warn? Contact Dermatitis, **28**: 163–165.

Abthoff J, Zahn W, Loose G, & Hirschmann A (1994) [Serial use of palladium for three-way-catalysts with high performance.] Motortech Z, **55**: 292–297 (in German).

Adachi A, Horikawa T, Takashima T, Komura T, Komura A, Tani M, & Ichihashi M (1997) Potential efficacy of low metal diets and dental metal elimination in the management of atopic dermatitis: an open clinical study. J Dermatol, **24**: 12–19.

Akerfeldt S & Lövgren G (1964) Spectrophotometric determination of disulfides, sulfinic acids, thio ethers, and thiols with the palladium(II) ion. Anal Biochem, **8**: 223–228.

Akiya O, Morimoto M, Suzuki Y, Hiura H, Katagiri S, & Kawashima Y (1996) A case of oral lichen planus due to sensitization to palladium. Bull Tokyo Dent Coll, **37**: 35–39.

Aldrich (1996) [Palladium and palladium compounds.] In: Katalog, Handbuch, Feinchemikalien 1996–1997. Steinheim, Sigma-Aldrich Chemie GmbH, pp 1262–1266 (in German).

Ando A & Ando I (1994) Biodistributions of radioactive bipositive metal ions in tumor-bearing animals. BioMetals, **7**: 185–192.

APHA, American Water Works Association, & Water Pollution Control Federation (1981) Bioassay methods for aquatic organisms. In: Franson MA ed. Standard methods for the examination of water and wastewater, 15th ed. Washington, DC, American Public Health Association, pp 612–742.

APHA, American Water Works Association, & Water Pollution Control Federation (1989) Metals by flame atomic absorption spectrometry. In: Franson MA ed. Standard methods for the examination of water and wastewater, 17th ed. Washington, DC, American Public Health Association, pp 3-13–3-28.

Aresta M, De Fazio M, Fumarulo R, Giordano D, Pantaleo R, & Riccardi S (1982) Biological activity of metal complexes: V. Influence of Pd(II), Pt(II), and Rh(I) on the macrophages chemotaxis. Biochem Biophys Res Commun, **104**: 121–125.

Artelt S (1997) [Search and identification of gaseous noble metal emissions from catalysts (VPT 06). Quantitative and statistical test studies on platinum aerosols emitted from vehicle catalysts (VPT 08).] In: Pohl D ed. [Noble metal emissions.] Neuherberge, GSF – National Research Centre for Environment and Health, pp 31–47 (in German).

Augthun M & Spiekermann H (1994) [*In vitro* and *in vivo* investigations on the corrosion behaviour of palladium alloy.] Dtsch Zahnaerztl Z, **49**: 632–635 (in German).

Augthun M, Lichtenstein M, & Kammerer G (1990) [Studies on the allergic potential of palladium alloys.] Dtsch Zahnaerztl Z, **45**: 480–482 (in German).

Augthun M, Kirkpatrick CJ, & Schyma S (1991) [Studies on the pneumoconiosis risk of dental laboratory technicians due to palladium particles in dust.] Dtsch Zahnaerztl Z, **46**: 519–522 (in German).

Baraj B, Sastre A, Martinez M, & Spahiu K (1996) Simultaneous determination of chloride complexes of Pt(IV) and Pd(II) by capillary zone electrophoresis with direct UV absorbance detection. Anal Chim Acta, **319**: 191–197.

Begerow J, Turfeld M, & Dunemann L (1997a) Determination of physiological noble metals in human urine using liquid–liquid extraction and Zeeman electrothermal atomic absorption spectrometry. Anal Chim Acta, **340**: 277–283.

Begerow J, Turfeld M, & Dunemann L (1997b) Determination of physiological palladium, platinum, iridium and gold levels in human blood using double focusing magnetic sector field inductively coupled plasma mass spectrometry. J Anal Atom Spectrom, **12**: 1095–1098.

Begerow J, Turfeld M, & Dunemann L (1997c) Determination of physiological palladium and platinum levels in urine using double focusing magnetic sector field ICP-MS. Fresenius J Anal Chem, **359**: 427–429.

Begerow J, Wiesmüller GA, Turfeld M, & Dunemann L (1998) [Contribution of platinum and palladium concentrations from road traffic to the background burden of these compounds in the general population.] Umweltmed Forsch Prax, **3**(4): 257 (in German).

Begerow J, Neuendorf J, Turfeld M, Raab W, & Dunemann L (1999a) Long-term urinary platinum, palladium, and gold excretion of patients after insertion of noble-metal dental alloys. Biomarkers, **4**(1): 27–36.

Begerow J, Sensen U, Wiesmüller GA, & Dunemann L (1999b) [Internal platinum, palladium, and gold exposure in environmentally and occupationally exposed persons.] Zentralbl Hyg Umweltmed, **202**(5): 411–424 (in German).

Benedict WG (1970) Some morphological and physiological effects of palladium on Kentucky bluegrass. Can J Bot, **48**: 91–93.

Bessing C & Kallus T (1987) Evaluation of tissue response to dental alloys by subcutaneous implantation. Acta Odontol Scand, **45**: 247–255.

BGA (1993) [Recommendations from 1st August 1993 for the reduction of risk concerning choice and further handling of dental inlays, fillings and orthodontic alloys.] In: Legierungen in der zahnaerztlichen Therapie. Bonn, Bundesgesundheitsamt, pp 5–18 (in German).

Biagini RE, Bernstein IL, Gallagher JS, Moorman WJ, Brooks S, & Gann PH (1985) The diversity of reaginic immune responses to platinum and palladium metallic salts. J Allergy Clin Immunol, **76**: 794–802.

Bikhazi AB, Salameh A, El-Kasti MM, & Awar RA (1995) Comparative nephrotoxic effects of *cis*-platinum (II), *cis*-palladium (II), and *cis*-rhodium (III) metal coordination compounds in rat kidneys. Comp Biochem Physiol, **111C**: 423–427.

Boman A & Wahlberg JE (1990) Experimental sensitization with palladium chloride in the guinea pig. Contact Dermatitis, **23**: 256.

Bonner FW & Parke DV (1984) [Transmissibility to humans of test results with laboratory animals.] In: Merian E ed. Metalle in der Umwelt. Weinheim, VCH Verlagsgesellschaft, pp 195–207 (in German).

Brasch J & Geier J (1997) Patch test results in schoolchildren. Results from the Information Network of Departments of Dermatology (IVDK) and the German Contact Dermatitis Research Group (DKG). Contact Dermatitis, **37**: 286–293.

Brenchley WE (1934) The effect of rubidium sulphate and palladium chloride on the growth of plants. Ann Appl Biol, **21**: 398–417.

Brockhaus (1970) [Saliva.] In: Stöcker FW & Dietrich G ed. Brockhaus ABC der Biologie. Leipzig, VEB F.A. Brockhaus Verlag, p 773 (in German).

Budavari S, O'Neil MJO, Smith A, Heckelman PE, & Kinneary JF ed. (1996) The Merck index, an encyclopedia of chemicals, drugs, and biologicals, 12th ed. Whitehouse Station, New Jersey, Merck & Co., p 1201.

Bünger J (1997) [The exhaust catalyst seen from the angle of environmental and occupational medicine.] Zentralbl Arbeitsmed Arbeitsschutz, **47**: 56–60 (in German).

Bünger J, Stork J, & Stalder K (1996) Cyto- and genotoxic effects of coordination complexes of platinum, palladium and rhodium *in vitro*. Int Arch Occup Environ Health, **69**: 33–38.

Camarasa JG & Serra-Baldrich E (1990) Palladium contact dermatitis. Am J Contact Dermatitis, **1**: 114–115.

Camarasa JG, Serra-Baldrich E, Lluch M, Malet A, & Calderon PG (1989) Recent unexplained patch test reactions to palladium. Contact Dermatitis, **20**: 388–389.

Camarasa JG, Burrows D, Menné T, Wilkinson JD, & Shaw S (1991) Palladium contact sensitivity. Contact Dermatitis, **24**: 370–371.

Campbell KI, George EL, Hall LL, & Stara JF (1975) Dermal irritancy of metal compounds. Arch Environ Health, **30**: 168–170.

Castan P, Colacio-Rodriguez E, Beauchamp AL, Cros S, & Wimmer S (1990) Platinum and palladium complexes of 3-methyl orotic acid: A route toward palladium complexes with good antitumor activity. J Inorg Biochem, **38**: 225–239.

Castelain PY & Castelain M (1987) Contact dermatitis to palladium. Contact Dermatitis, **16**: 46.

Cederbrant K, Hultman P, Marcusson JA, & Tibbling L (1997) *In vitro* lymphocyte proliferation as compared to patch test using gold, palladium and nickel. Int Arch Allergy Immunol, **112**: 212–217.

Chen M, Gleichmann E, & Mikecz AV (1998) Altered subcellular localization of scleroderma autoantigen fibrillarin by heavy metals. Naunyn-Schmiedeberg's Arch Pharmacol, **357** (Suppl 4): R128.

Chikuma M, Aoki H, & Tanaka H (1991) Determination of metal ions in environmental waters by flameless atomic absorption spectrometry combined with preconcentration using a sulfonated dithizone-loaded resin. Anal Sci, Suppl **7**: 1131–1134.

Chiu DTY & Liu TZ (1997) Free radical and oxidative damage in human blood cells. J Biomed Sci, **4**: 256–259.

Christensen GM (1971/72) Effects of metal cations and other chemicals upon the *in vitro* activity of two enzymes in the blood plasma of the white sucker, *Catostomus commersoni* (Lacépède). Chem-Biol Interact, **4**: 351–361.

Christensen GM & Olson DL (1981) Effect of water pollutants and other chemicals upon ribonuclease activity *in vitro*. Environ Res, **26**: 274–280.

Clothier RH, Hulme L, Ahmed AB, Reeves HL, Smith M, & Balls M (1988) *In vitro* cytotoxicity of 150 chemicals to 3T3-L1 cells, assessed by the FRAME kenacid blue method. ATLA — Altern Lab Anim, **16**: 84–95.

Coombes JS (1990) Palladium. In: Platinum 1990. London, Johnson Matthey plc, pp 51–57 (internal report).

Cotton FA & Wilkinson G (1982) [Platinum metals.] In: Cotton FA & Wilkinson G ed. [Inorganic chemistry: an advanced summary], 4th ed. Weinheim, Verlag Chemie, pp 917–981 (in German).

Cowley A (1997) Palladium. In: Platinum 1997. London, Johnson Matthey plc, pp 35–37, 50–51.

Cowley A (1998) Palladium. In: Platinum 1998. London, Johnson Matthey plc, pp 33–37, 50–51.

Cowley A (1999) Palladium. In: Platinum 1999. London, Johnson Matthey plc, pp 31–37, 50–51.

Cubelic M, Pecoroni R, Schäfer J, Eckhardt JD, Berner Z, & Stüben D (1997) Verteilung verkehrsbedingter Edelmetallimmissionen in Böden. Umweltwiss Schadstoff-Forsch, **9**: 249–258.

Culliton CR, Meenaghan MA, Sorensen SE, Greene GW, & Eick JD (1981) A critical evaluation of the acute systemic toxicity test for dental alloys using histopathologic criteria. J Biomed Mater Res, **15**: 565–575.

Curic M, Tusek-Bozic L, Vikic-Topic D, Scarcia V, Furlani A, Balzarini J, & De Clercq E (1996) Palladium(II) complexes of dialkyl alpha-anilinobenzylphosphonates. Synthesis, characterization, and cytostatic activity. J Inorg Biochem, **63**: 125–142.

Daenen M, Rogiers P, Van de Walle C, Rochette F, Demedts M, & Nemery B (1997) Occupational asthma caused by exposure to palladium. Eur Respir J, **10** (Suppl 25): 165s.

Daenen M, Rogiers P, Van de Walle C, Rochette F, Demedts M, & Nemery B (1999) Occupational asthma caused by palladium. Eur Respir J, **13**(1): 213–216.

Daunderer M (1993) [Toxicological information on single compounds, palladium III-3.] In: Daunderer M ed. Handbuch der Umweltgifte: Klinische Umwelttoxikologie für die Praxis, 7th ed. Landsberg, Ecomed, pp 1–67 (in German).

Daunderer M (1994) [Toxicological information on single compounds, metals III-3]. In: Daunderer M ed. Handbuch der Umweltgifte: Klinische Umwelttoxikologie für die Praxis, 14th ed. Landsberg, Ecomed, pp 1–41 (in German).

De Fine Olivarius F & Menné T (1992) Contact dermatitis from metallic palladium in patients reacting to palladium chloride. Contact Dermatitis, **27**: 71–73.

Degussa (1997) Hannover Messe '97: Special display automobile recycling, electronics recycling. The Degussa catalytic converter association (Kat-Verbund). Hanover, Deutsche Messe AG, 1997 (Internet communication of 9 September 1997 at web site http://pluto.messe.de:8000/hm97/sonderschau/autoe.html).

Degussa AG (ed) (1995) Edelmetall-Taschenbuch, 2nd ed. Heidelberg, Hüthig-Verlag.

Deichmann WB & LeBlanc TJ (1943) Determination of the approximate lethal dose with about six animals. J Ind Hyg Toxicol, **25**: 415–417.

Dissanayake CB, Kritsotakis K, & Tobschall HJ (1984) The abundance of Au, Pt, Pd, and the mode of heavy metal fixation in highly polluted sediments from the Rhine River near Mainz, West Germany. Int J Environ Stud, **22**: 109–119.

Doudoroff P & Katz M (1953) Critical review of literature on the toxicity of industrial wastes and their components to fish. II. The metals, as salts. Sewage Ind Wastes, **25**: 802–839.

Downey D (1989) Contact mucositis due to palladium. Contact Dermatitis, **21**: 54.

Downey D (1992) Porphyria induced by palladium–copper dental prostheses: a clinical report. J Prosthet Dent, **67**: 5–6.

Drápal S & Pomajbík J (1993) Segregation in precious-metal dental-casting alloys. J Dent Res, **72**: 587–591.

Durbin PW (1960) Metabolic characteristics within a chemical family. Health Phys, **2**: 225–238.

Durbin PW, Scott KG, & Hamilton JG (1957) The distribution of radioisotopes of some heavy metals in the rat. Univ Calif Berkeley Publ Pharmacol, 3(1): 1–27.

Eedy DJ, Burrows D, & McMaster D (1991) The nickel content of certain commercially available metallic patch test materials and its relevance in nickel-sensitive subjects. Contact Dermatitis, **24**: 11–15.

Eimerl S & Schramm M (1993) Potentiation of ^{45}Ca uptake and acute toxicity mediated by the N-methyl-D-aspartate receptor: the effect of metal binding agents and transition metal ions. J Neurochem, **61**: 518–525.

Eisenring R, Wirz J, Rahn BA, & Geret V (1986) [Biological examination of metalloceramic alloys.] Schweiz Monatsschr Zahnmed, **96**: 500–520 (in German).

Eller R, Alt F, Tölg G, & Tobschall HJ (1989) An efficient combined procedure for the extreme trace analysis of gold, platinum, palladium and rhodium with the aid of graphite furnace atomic absorption spectrometry and total-reflection X-ray fluorescence analysis. Fresenius Z Anal Chem, **334**: 723–739.

Eneström S & Hultman P (1995) Does amalgam affect the immune system? A controversial issue. Int Arch Allergy Immunol, **106**: 180–203.

Estler C-J (1992) [How toxic is palladium?] Dtsch Zahnaerztl Z, **47**: 361–363 (in German).

Fabri J, Dabelstein W, & Reglitzky A (1990) Motor fuels. In: Elvers B, Hawkins S, & Schulz G ed. Ullmann's encyclopedia of industrial chemistry, 5th ed. Vol. A16. Weinheim, VCH Verlagsgesellschaft, pp 719–753.

Farago ME & Parsons PJ (1985a) The recovery of platinum metals by the water hyacinth. Environ Technol Lett, **6**: 165–174.

Farago ME & Parsons PJ (1985b) Effects of platinum metals on plants. Trace Subst Environ Health, **19**: 397–407.

Farago ME & Parsons PJ (1994) The effects of various platinum metal species on the water plant *Eichhornia crassipes* (MART.) Solms. Chem Spec Bioavailab, **6**: 1–12.

Fawwaz RA, Hemphill W, & Winchell HS (1971) Potential use of [109]Pd–porphyrin complexes for selective lymphatic ablation. J Nucl Med, **12**: 231–236.

Fernandez-Redondo V, Gomez-Centeno P, & Toribio J (1998) Chronic urticaria from a dental bridge. Contact Dermatitis, **38**: 178.

Finger PT, Berson A, & Szechter A (1999) Palladium-103 plaque radiotherapy for choroidal melanoma. Ophthalmology, **106**(3): 606–613.

Fishbein L (1976) Potential impact of newer materials: Noble metals, flame retardants, new pest-control strategies. Research Triangle Park, North Carolina, National Institutes of Health, National Institute of Environmental Health Sciences (NIH/NIEHS-77/008).

Fisher RF, Holbrook DJ Jr, Leake HB, & Brubaker PE (1975) Effect of platinum and palladium salts on thymidine incorporation into DNA of rat tissues. Environ Health Perspect, **12**: 57–62.

Flint GN (1998) A metallurgical approach to metal contact dermatitis. Contact Dermatitis, **39**: 213–221.

Fothergill SJR, Withers DF, & Clements FS (1945) Determination of traces of platinum and palladium in the atmosphere of a platinum refinery. Br J Ind Med, **2**: 99–101.

Fowler JF (1990) Allergic contact dermatitis to metals. Am J Contact Dermatitis, **1**: 212–223.

Freiesleben D, Wagner B, Hartl H, Beck W, Hollstein M, & Lux F (1993) [Solution of palladium and platinum powder through biogenic compounds.] Z Naturforsch, **48B**: 847–848 (in German).

Fuchs WA & Rose AW (1974) The geochemical behavior of platinum and palladium in the weathering cycle in the stillwater complex, Montana. Econ Geol, **69**: 332–346.

Fuhrmann R (1992) [Subcutaneous oral toxicity of dental alloys — an experimental animal study.] Marburg, Philipps-University (thesis) (in German).

Fujita S (1971) [Experimental studies on carcinogenicity of physical stimuli.] Shika Igaku, **34**: 918–932 (in Japanese).

Gailhofer G & Ludvan M (1990) The significance of positive patch test reactions to palladium. Allergologie, **13**: 250.

Gebel T, Lantzsch H, Pleßow K, & Dunkelberg H (1997) Genotoxicity of platinum and palladium compounds in human and bacterial cells. Mutat Res, **389**: 183–190.

Geldmacher-von Mallinckrodt M & Pooth M (1969) [Simultaneous spectrographic testing of 25 metals and metalloids in biological materials.] Arch Toxicol, **25**: 5–18 (in German).

German Federal Environment Agency (1997) [Commission on Human Biomonitoring: "Saliva test" — mercury exposure due to amalgam fillings.] Bundesgesundheitsblatt, **2**: 76 (in German).

Gertler AW (1994) A preliminary apportionment of the sources of fine particulates impacting on the Israeli coast. Isr J Chem, **34**: 425–433.

Gofman JW, deLalla OF, Kovich EL, Lowe O, Martin W, Piluso DL, Tandy RK, & Upham F (1964) Chemical elements of blood of man. Arch Environ Health, **8**: 105–109.

Goldberg ED (1987) Heavy metal analyses in the marine environment — approaches to quality control. Mar Chem, **22**: 117–124.

Goldberg ED, Koide M, Yang JS, & Bertine KK (1988) Comparative marine chemistry of the platinum group metals and their periodic table neighbors. In: Kramer JR & Allen HE ed. Metal speciation: Theory, analysis and application. Chelsea, Michigan, Lewis Publishers, pp 201–217.

Gomez MB, Gomez MM, & Palacios MA (2000) Control of interferences in the determination of Pt, Pd and Rh in airborne particulate matter by inductively coupled plasma mass spectroscopy. Anal Chim Acta, **404**: 285–294.

Graham RD & Williams DR (1979) The synthesis and screening for anti-bacterial, -cancer, -fungicidal and -viral activities of some complexes of palladium and nickel. J Inorg Nucl Chem, **41**: 1245–1249.

Grant WM & Schuman JS (1993) Toxicology of the eye: Effects on the eyes and visual system from chemicals, drugs, metals and minerals, plant toxins and venoms; also, systemic side effects from eye medications, 4th ed. Springfield, Illinois, Thomas Books, pp 1099, 1374.

Griem P & Gleichmann E (1995) Metal ion induced autoimmunity. Curr Opin Immunol, **7**: 831–838.

Grill V, Sandrucci MA, Basa M, Di Lenarda R, Dorigo E, Narducci P, Martelli AM, Delbello G, & Bareggi R (1997) The influence of dental metal alloys on cell proliferation and fibronectin arrangement in human fibroblast cultures. Arch Oral Biol, **42**(9): 641–647.

Guerra L, Misciali C, Borello P, & Melino M (1988) Sensitization to palladium. Contact Dermatitis, **19**: 306–307.

Guidetti L & Stefanetti M (1996) [Biomonitoring of atmospheric deposition of trace elements using the lichen *Parmelia caperata* in the area surrounding Lake Orta.] Acqua Aria, **5**: 489–497 (in Italian).

Hackel H, Miller K, Elsner P, & Burg G (1991) Short communications: Unusual combined sensitization to palladium and other metals. Contact Dermatitis, **24**: 131–157.

Hall GEM & Pelchat JC (1993) Determination of palladium and platinum in fresh waters by inductively coupled plasma mass spectrometry and activated charcoal preconcentration. J Anal Atom Spectrom, **8**: 1059–1065.

Hansen PA & West LA (1997) Allergic reaction following insertion of a Pd–Cu–Au fixed partial denture: A clinical report. J Prosthodont, **6**(2): 144–148.

Hay C & Ormerod AD (1998) Severe oral and facial reaction to 6 metals in restorative dentistry. Contact Dermatitis, **38**: 216.

Hees T, Wenclawiak B, Lustig S, Schramel P, Schwarzer M, Schuster M, Verstraete D, Dams R, & Helmers E (1998) Distribution of platinum group elements (Pt, Pd, Rh) in environmental and clinical matrices: Composition, analytical techniques and scientific outlook. Environ Sci Pollut Res Int, **5**(2): 105–111.

Helmers E & Kümmerer K (1997) Series: Platinum group elements in the environment — anthropogenic impact. Environ Sci Pollut Res, **4**(2): 99–103.

Helmers E, Schwarzer M, & Schuster M (1998) Comparison of palladium and platinum in environmental matrices polluted by automobile emissions. Environ Sci Pollut Res, **5**(1): 44–50.

Herblin WF & Ritt PE (1964) Short communications: The binding of heavy metal ions by papain. Biochem Biophys Acta, **85**: 489–490.

Higgins JD, Neely L, & Fricker S (1993) Synthesis and cytotoxicity of some cyclometallated palladium complexes. J Inorg Biochem, **49**: 149–156.

Hill RF & Mayer WJ (1977) Radiometric determination of platinum and palladium attrition from automotive catalysts. IEEE Trans Nucl Sci, **24**: 2549–2554.

Hodge VF & Stallard MO (1986) Platinum and palladium in roadside dust. Environ Sci Technol, **20**: 1058–1060.

Holbrook DJ Jr (1977) Content of platinum and palladium in rat tissue: Correlation of tissue concentration of platinum and palladium with biochemical effects. Research Triangle Park, North Carolina, US Environmental Protection Agency (EPA-600/1-77-051).

Holbrook DJ Jr, Washington ME, Leake HB, & Brubaker PE (1975) Studies on the evaluation of the toxicity of various salts of lead, manganese, platinum, and palladium. Environ Health Perspect, **10**: 95–101.

Holbrook DJ Jr, Washington ME, Leake HB, & Brubaker PE (1976) Effects of platinum and palladium salts on parameters of drug metabolism in rat liver. J Toxicol Environ Health, **1**: 1067–1079.

Holleman AF & Wiberg E (1995) [Palladium and platinum.] In: Wiberg N ed. Lehrbuch der Anorganischen Chemie, 101st ed. Berlin, De Gruyter, pp 1587–1604 (in German).

Howarth F & Cooper ERA (1955) The fate of certain foreign colloids and crystalloids after subarachnoid injection. Acta Anat, **25**: 112–140.

HSE (1983) The medical monitoring of workers exposed to platinum "salts." Health and Safety Executive. London, Her Majesty's Stationery Office, 2 pp (MS/22).

Hueso-Urena F, Moreno-Carretero MN, Salas-Peregrín JM, & Alvarez de Cienfuegos-López G (1991) Palladium, platinum, cadmium, and mercury complexes with neutral isoorotic and 2-thioisoorotic acids: IR and NMR spectroscopies, thermal behavior and biological properties. J Inorg Biochem, **43**: 17–27.

Hussain MZ, Bhatnagar RS, & Lee SD (1977) Biochemical mechanisms of interaction of environmental metal contaminants with lung connective tissue. In: Lee SD ed. Biochemical effects of environmental pollutants. Ann Arbor, Michigan, Ann Arbor Science, pp 341–350.

Hysell D, Neiheisel S, & Cmehil D (1974) Ocular irritation of two palladium and two platinum compounds in rabbits. In: Studies on catalytic components and exhaust emissions. Cincinnati, Ohio, US Environmental Protection Agency, Environmental Toxicological Research Laboratory, National Environmental Research Center [cited in NAS, 1977].

Inacker O & Malessa R (1997) [Experimental study on the transfer of platinum from vehicle exhaust catalysts (VPO 03).] In: Pohl D ed. Noble metal emissions. Neuherberge, GSF – National Research Centre for Environment and Health, pp 48–67 (in German).

IPCS (1991) Environmental Health Criteria 125: Platinum. Geneva, World Health Organization, International Programme on Chemical Safety, 167 pp.

IPCS (1999) Environmental Health Criteria 212: Principles and methods for assessing allergic hypersensitization associated with exposure to chemicals. Geneva, World Health Organization, International Programme on Chemical Safety, 399 pp.

Ito A, Okazaki Y, Tateishi T, & Ito Y (1995) In vitro biocompatibility, mechanical properties, and corrosion resistance of Ti–Zr–Nb–Ta–Pd and Ti–Sn–Nb–Ta–Pd alloys. J Biomed Mater Res, **29**: 893–900.

Jain N, Mittal R, Srivastava TS, Satyamoorthy K, & Chitnis MP (1994) Synthesis, characterization, DNA binding, and cytotoxic studies of dinuclear complexes of palladium(II) and platinum(II) with 2,2-bipyridine and alpha,omega-diaminoalkane-N,N'-diacetic acid. J Inorg Biochem, **53**: 79–94.

Jappe U, Bonnekoh B, & Gollnik H (1999) Persistent granulomatous contact dermatitis due to palladium body-piercing ornaments. Contact Dermatitis, **40**: 111–112.

Johnson DE, Tillery JB, & Prevost RJ (1975a) Trace metals in occupationally and nonoccupationally exposed individuals. Environ Health Perspect, **10**: 151–158.

Johnson DE, Tillery JB, & Prevost RJ (1975b) Levels of platinum, palladium, and lead in populations of southern California. Environ Health Perspect, **12**: 27–33.

Johnson DE, Prevost RJ, Tillery JB, Caman DE, Hosenfeld JM (1976) Baseline levels of platinum and palladium in human tissue. Research Triangle Park, North Carolina, US Environmental Protection Agency (EPA-600/1-76-019).

Johnson Matthey (1994a) Palladium II chloride: Acute oral toxicity (limit test) in the rat. Berkshire, Johnson Matthey Technology Centre (Project No. 52/56; unpublished report).

Johnson Matthey (1994b) Palladium II chloride: Acute dermal irritation test in the rabbit. Berkshire, Johnson Matthey Research Centre (Project No. 52/57; unpublished report).

Johnson Matthey (1995a) Tetraammine palladium hydrogen carbonate: Acute oral toxicity test in the rat. Hertfordshire, Johnson Matthey plc (SPL Project No. 036/045; unpublished report).

Johnson Matthey (1995b) Tetraammine palladium hydrogen carbonate: Acute dermal irritation test in the rabbit. Hertfordshire, Johnson Matthey plc (SPL Project No. 036/046; unpublished report).

Johnson Matthey (1995c) Tetraammine palladium hydrogen carbonate: Acute eye irritation test in the rabbit. Hertfordshire, Johnson Matthey plc (SPL Project No. 036/047; unpublished report).

Johnson Matthey (1995d) Tetraammine palladium hydrogen carbonate: Reverse mutation assay "Ames test" using *Salmonella typhimurium*. Hertfordshire, Johnson Matthey plc (SPL Project No. 036/049; unpublished report).

Johnson Matthey (1995e) Tetraammine palladium hydrogen carbonate: Assessment of the inhibitory effect on the respiration of activated sewage sludge. Hertfordshire, Johnson Matthey plc (SPL Project No. 036/050; unpublished report).

Johnson Matthey (1997a) Tetraammine palladium hydrogen carbonate: Acute dermal toxicity (limit test) in the rat. Hertfordshire, Johnson Matthey plc (SPL Project No. 036/083; unpublished report).

Johnson Matthey (1997b) Tetraammine palladium hydrogen carbonate: Twenty-eight day repeated dose oral (gavage) toxicity study in the rat. Hertfordshire, Johnson Matthey plc (SPL Project No. 036/084; unpublished report).

Johnson Matthey (1997c) Tetraammine palladium hydrogen carbonate: Magnusson & Kligman maximisation study in the guinea pig. Hertfordshire, Johnson Matthey plc (SPL Project No. 036/048; unpublished report).

Johnson Matthey (1997d) Tetraammine palladium hydrogen carbonate: Chromosome aberration test in human lymphocytes *in vitro*. Conducted by Safepharm Laboratories Ltd, Derby, for Johnson Matthey plc, Hertfordshire (SPL Project No. 036/085; unpublished report).

Johnson Matthey (1997e) Tetraammine palladium hydrogen carbonate: Algal inhibition test. Hertfordshire, Johnson Matthey plc (SPL Project No. 036/088; unpublished report).

Johnson Matthey (1997f) Tetraammine palladium hydrogen carbonate: Acute toxicity to *Daphnia magna*. Hertfordshire, Johnson Matthey plc (SPL Project No. 036/087; unpublished report).

Johnson Matthey (1997g) Tetraammine palladium hydrogen carbonate: Acute toxicity to rainbow trout (*Oncorhynchus mykiss*). Hertfordshire, Johnson Matthey plc (SPL Project No. 036/086; unpublished report).

Johnson Matthey (1998) Tetraammine palladium hydrogen carbonate: Micronucleus test in the mouse. Hertfordshire, Johnson Matthey plc (SPL Project No. 036/104; unpublished report).

Johnson Matthey (2000) Material safety data sheet: Tetraammine palladium hydrogen carbonate. Berkshire, Johnson Matthey Technology Centre.

Jones AH (1976) Determination of platinum and palladium in blood and urine by flameless atomic absorption spectrophotometry. Anal Chem, **48**: 1472–1474.

Jones EA, Warshawsky A, Dixon K, Nicolas DJ, & Steele TW (1977) The group extraction of noble metals with s-(1-decyl)-*N,N'*-diphenylisothiouronium bromide and their determination in the organic extract by atomic-absorption spectrometry. Anal Chim Acta, **94**: 257–268.

Jones MM, Schoenheit JE, & Weaver AD (1979) Pretreatment and heavy metal LD50 values. Toxicol Appl Pharmacol, **49**: 41–44.

Kamboj VP & Kar AB (1964) Antitesticular effect of metallic and rare earth salts. J Reprod Fertil, **7**: 21–28.

Kanerva L, Kerosuo H, Kullaa A, & Kerosuo E (1996) Allergic patch test reactions to palladium chloride in schoolchildren. Contact Dermatitis, **34**: 39–42.

Kansu G & Aydin AK (1996) Evaluation of the biocompatibility of various dental alloys: Part I — Toxic potentials. Eur J Prosthodont Restor Dent, 4(3): 129–136.

Katoh N, Hirano S, Kishimoto S, & Yasuno H (1999) Dermal contact dermatitis caused by allergy to palladium. Contact Dermatitis, 40: 226–227.

Kauffmann M (1913) [A new degreasing agent: colloid palladium hydroxydul ("Leptynol").] Münch Med Wochenschr, 60: 525–527 (in German).

Kawahara H, Yamagami A, & Nakamura M (1968) Biological testing of dental materials by means of tissue culture. Int Dent J, 18: 443–467.

Kawata Y, Shiota M, Tsutsui H, Yoshida Y, Sasaki H, & Kinouchi Y (1981) Cytotoxicity of Pd–Co dental casting ferromagnetic alloys. J Dent Res, 60: 1403–1409.

Kenawy IM, Khalifa ME, & El-Defrawy MM (1987) Preconcentration and determination (AAS) of trace Ag(I), Au(III), Pd(II) and Pt(IV) using a cellulose ion-exchange (Hyphan). Analusis, 15: 314–317.

Khan BT, Bhatt J, Najmuddin K, Shamsuddin S, & Annapoorna K (1991) Synthesis, antimicrobial, and antitumor activity of a series of palladium (II) mixed ligand complexes. J Inorg Biochem, 44: 55–63.

Khangarot BS (1991) Toxicity of metals to a freshwater tubificid worm, Tubifex tubifex (Muller). Bull Environ Contam Toxicol, 46: 906–912.

Kobayashi H (1989) [Study of the influence of the anodic potential on metal-components dissolution from dental alloys.] Shikwa Gakuho, 89: 1679–1697 (in Japanese).

Koch P & Bahmer FA (1995) Oral lichenoid lesions, mercury hypersensitivity and combined hypersensitivity to mercury and other metals: histologically-proven reproduction of the reaction by patch testing with metal salts. Contact Dermatitis, 33: 323–328.

Koch P & Bahmer FA (1999) Oral lesions and symptoms related to metals used in dental restorations: a clinical, allergological, and histologic study. J Am Acad Dermatol, 41(3): 422–430.

Koch P & Baum HP (1996) Contact stomatitis due to palladium and platinum in dental alloys. Contact Dermatitis, 34: 253–257.

Koch RC & Roesmer J (1962) Application of activation analysis to the determination of trace-element concentrations in meat. J Food Sci, 27: 309–320.

Kolesova GM, Zakharova IA, Raykhman LM, & Moshkovski YS (1979) [Effect of platinum and palladium complexes on enzymatic system of mitochondria.] Vopr Med Khim, 25: 537–540 (in Russian).

Kolpakov FI, Kolpakova AF, & Prochorenkov VI (1980) [Toxic and sensitizing properties of palladium hydrochloride.] Gig Tr Prof Zabol, 4: 52–54 (in Russian).

Kothny EL (1979) Palladium in plant ash. Plant Soil, 53: 547–550.

Kozlowski H & Pettit LD (1991) Amino acid and peptide complexes of the platinum group metals. In: Hartley FR ed. Chemistry of the platinum group metals: recent developments. New York, Elsevier, pp 530–545.

Kränke B & Aberer W (1996) Multiple sensitivities to metals. Contact Dermatitis, **34**: 225.

Kränke B, Binder M, Derhaschnig J, Komericki P, Pirkhammer D, Ziegler V, & Aberer W (1995) [Testing with the "Austrian standard patch test series" — test epidemiologic data and results.] Wien Klin Wochenschr, **107**: 323–330 (in German).

Kratzenstein B & Weber H (1988) [Metallic dental replacement materials in sensibilized patients.] Dtsch Zahnaerztl Z, **43**: 419–423 (in German).

Kratzenstein B, Sauer KH, & Weber H (1988) [*In vivo* corrosion symptoms from cast restorations and their reciprocal effects in the cavity of the mouth.] Dtsch Zahnaerztl Z, **43**: 343–348 (in German).

Kroschwitz JI ed. (1996) Platinum-group metals. In: Kirk-Othmer — Encyclopedia of chemical technology. New York, John Wiley & Sons, pp 347–406.

Kulig J, Pagels J, Wiesenborn A, Gleichmann E, Kind P, & Schuppe HC (1995) Palladium salts are immunogenic in mice. Arch Dermatol Res, **287**: 389.

Kulig J, Rönnau A, Sachs B, Schürer NY, & Schuppe HC (1997) Contact hypersensitivity to palladium salts: Demonstration of specific lymphocyte reactivity *in vitro*. Arch Dermatol Res, **289**: A44.

Kump LR & Byrne RH (1989) Palladium chemistry in seawater. Environ Sci Technol, **23**(6): 663–665.

Kütting B & Brehler R (1994) Klinisch relevante solitäre Palladiumallergie. Hautarzt, **45**: 176–178.

Lantzsch H & Gebel T (1997) Genotoxicity of selected metal compounds in the SOS chromotest. Mutat Res, **389**: 191–197.

Lassig JP, Shultz MD, Gooch MG, Evans BR, & Woodward J (1995) Inhibition of cellobiohydrolase I from *Trichoderma reesei* by palladium. Arch Biochem Biophys, **322**: 119–126.

Lee DS (1983) Palladium and nickel in north-east Pacific waters. Nature, **305**: 47–48.

Lee K-I, Tashiro T, & Noji M (1994) Platinum and palladium complexes containing ethylene-diamine derivatives as carrier ligands and their antitumor activity. Chem Pharm Bull, **42**: 702–703.

Li J-H & Byrne RH (1990) Amino acid complexation of palladium in seawater. Environ Sci Technol, **24**: 1038–1041.

Lide DR ed. (1992) The elements. In: CRC handbook of chemistry and physics: A ready-reference book of chemical and physical data, 72nd ed. Boston, Massachusetts, CRC Press.

Liden C & Wahlberg JE (1994) Cross-reactivity to metal compounds studied in guinea pigs induced with chromate or cobalt. Acta Derm Venereol, **74**: 341–343.

Lin H-X, Li Z-L, Dai G-L, Bi Q-S, & Yu R-Q (1993) Preliminary fluorimetric screening of fourteen palladium complexes as potential antitumor agents. Sci China B, **36**: 1216–1223.

Litchfield IJ Jr & Wilcoxon FI (1949) A simplified method for evaluating dose–effect experiments. J Pharmacol Exp Ther, **96**: 99–113.

Liu TZ, Lee SD, & Bhatnagar RS (1979a) Toxicity of palladium. Toxicol Lett, **4**: 469–473.

Liu TZ, Khayam-Bashi H, & Bhatnagar RS (1979b) Inhibition of creatine kinase activity and alterations in electrophoretic mobility by palladium ions. J Environ Pathol Toxicol, **2**: 907–916.

Liu TZ, Chou LY, & Humphreys MH (1979c) Inhibition of intestinal alkaline phosphatase by palladium. Toxicol Lett, **4**: 433–438.

Liu TZ, Lin TF, Chiu DTY, Tsai K-J, & Stern A (1997) Palladium or platinum exacerbates hydroxyl radical mediated DNA damage. Free Radical Biol Med, **23**: 155–161.

Loebenstein RJ (1996) Statistical compendium — Platinum-group metals. Reston, Virginia, US Geological Survey Minerals Information (Internet communication of 17 March 1997 at web site http://minerals.er.usgs.gov/minerals/pubs/commodity/platinum/stat).

Lottermoser BG (1995) Noble metals in municipal sewage sludges of southeastern Australia. Ambio, **24**: 354–357.

Lowenthal DH, Zielinska B, Chow JC, Watson JG, Gautam M, Ferguson DH, Neuroth GR, & Stevens KD (1994) Characterization of heavy-duty diesel vehicle emissions. Atmos Environ, **28**: 731–743.

Lu Z, Chow J, Watson J, Frazier C, Pritchett L, Dippel W, Bates B, Jones W, Torres G, Fisher R, & Lam D (1994) Temporal and spatial variations of PM_{10} aerosol in the Imperial Valley/Mexicali air quality study. Proc Annu Meet Air Waste Manage Assoc, **7**: 1–17.

Lüdke C, Hoffmann E, Skole J, & Artelt S (1996) Particle analysis of car exhaust by ETV-ICP-MS. Fresenius J Anal Chem, **355**: 261–263.

MAFF (1997) FSIS 131, November 1997: 1994 Total Diet Study: metals and other elements. London, Ministry of Agriculture, Fisheries and Food, Joint Food Safety and Standards Group (Internet communication of 14 January 2000 at web site http://www.maff.gov.uk/food/infsheet/ 1997/no131/131tds.htm).

MAFF (1998) FSIS 149, May 1998: 1994 Total Diet Study (Part 2) — Dietary intakes of metals and other elements. London, Ministry of Agriculture, Fisheries and Food, Joint Food Safety and Standards Group (Internet communication of 13 January 2000 at web site http://www.maff.gov.uk/ food/infsheet/1998/no149/149tds.htm).

Magnusson B & Kligman AM (1970) Allergic contact dermatitis in the guinea pig: identifications of contact allergens. Springfield, Illinois, Thomas Books, 141 pp.

Mansuri-Torshizi H, Mital R, Srivastava TS, Parekh H, & Chitnis MP (1991) Synthesis, characterization, and cytotoxic studies of alpha-diimine/1,2-diamine platinum(II) and palladium(II) complexes of selenite and tellurite and binding of some of these complexes to DNA. J Inorg Biochem, **44**: 239–247.

Mansuri-Torshizi H, Srivastava TS, Parekh HK, & Chitnis MP (1992a) Synthesis, spectroscopic, cytotoxic, and DNA binding studies of binuclear 2,2'-bipyridine-platinum(II) and -palladium(II) complexes of *meso*-alpha, alpha'-diaminoadipic and *meso*-alpha, alpha'-diaminosuberic acids. J Inorg Biochem, **45**: 135–148.

Mansuri-Torshizi H, Srivastava TS, Chavan SJ, & Chitnis MP (1992b) Cytotoxicity and DNA binding studies of several platinum (II) and palladium (II) complexes of the 2,2'-bipyridine and an anion of 2-pyridinecarboxylic/2-pyrazinecarboxylic acid. J Inorg Biochem, **48**: 63–70.

Marcusson JA (1996) Contact allergies to nickel sulfate, gold sodium thiosulfate and palladium chloride in patients claiming side-effects from dental alloy components. Contact Dermatitis, **34**: 320–323.

Marx H (1987) [The present situation concerning palladium alloys.] Zahnaerztl Mitt, **3**: 211–224 (in German).

Matilla A, Tercero JM, Dung N-H, Viossat B, Pérez JM, Alonso C, Martín-Ramos JD, & Niclós-Gutiérrez J (1994) *Cis*-Dichloro-palladium(II) complexes with diaminosuccinic acid and its diethyl ester: synthesis, molecular structure, and preliminary DNA-binding and antitumor studies. J Inorg Biochem, **55**: 235–247.

Matusiewicz H & Barnes RM (1988) Determination of metal chemotherapeutic agents in human body fluids using inductively coupled plasma atomic emission spectrometry with electrothermal vaporization. Acta Chim Hung, **125**: 777–784.

Mayer T (1989) [Allergic reactions to gold and palladium dental alloys in the epicutaneous test.] Cologne, University of Cologne (thesis) (in German).

McDonald I, Hart RJ, & Tredoux M (1994) Determination of the platinum-group elements in South African kimberlites by nickel sulphide fire-assay and neutron activation analysis. Anal Chim Acta, **289**: 237–247.

Meek SF, Harrold GC, & McCord CP (1943) The physiologic properties of palladium. Ind Med, **12**: 447–448.

Menck HR & Henderson BE (1976) Occupational differences in rates of lung cancer. J Occup Med, **18**: 797–801.

Merget R (1991) [Asthma bronchiale through metal compounds.] Frankfurt/Main, Johann Wolfgang Goethe University (professorial dissertation) (in German).

Minoia C, Sabbioni E, Apostoli P, Pietra R, Pozzoli L, Gallorini M, Nicolaou G, Alessio L, & Capodaglio E (1990) Trace element reference values in tissues from inhabitants of the European Community: I. A study of 46 elements in urine, blood and serum of Italian subjects. Sci Total Environ, **95**: 89–105.

Mital R, Srivastava TS, Parekh HK, & Chitnis MP (1991) Synthesis, characterization, DNA binding, and cytotoxic studies of some mixed-ligand palladium (II) and platinum (II) complexes of alpha-diimine and amino acids. J Inorg Biochem, **41**: 93–103.

Mital R, Shah GM, Srivastava TS, & Bhattacharya RK (1992) The effect of some new platinum (II) and palladium (II) coordination complexes on rat hepatic nuclear transcription *in vitro*. Life Sci, **50**: 781–790.

Mizoguchi S, Setoyama M, & Kanzaki T (1998) Linear lichen planus in the region of the mandibular nerve caused by an allergy to palladium in dental metals. Dermatology, **196**: 268–270.

Mjör IA & Christensen GJ (1993) Assessment of local side effects of casting alloys. Quintessence Int, **24**: 343–351.

MMAJ (1999) Metal Mining Agency of Japan (Internet communication of 5 May 2000 at web site http://www.mmaj.go.jp/mmaj_e/home.html).

Moldovan M, Gómez MM, & Palacios MA (1999) Determination of platinum, rhodium and palladium in car exhaust fumes. J Anal Atom Spectrom, 14: 1163–1169.

Moore W Jr, Hysell D, Crocker W, & Stara J (1974) Biological fate of [103]Pd in rats following different routes of exposure. Environ Res, 8: 234–240.

Moore W, Hysell D, Hall L, Campbell K, & Stara J (1975) Preliminary studies on the toxicity and metabolism of palladium and platinum. Environ Health Perspect, 10: 63–71.

Morgan JJ & Stumm W (1991) Chemical processes in the environment, relevance of chemical speciation. In: Merian E ed. Metals and their compounds in the environment — occurrence, analysis and biological relevance. Weinheim, VCH Verlagsgesellschaft, pp 67–103.

Moulon C, Vollmer J, & Weltzien HU (1995) Characterization of processing requirements and metal cross-reactivities in T cell clones from patients with allergic contact dermatitis to nickel. Eur J Immunol, 25: 3308–3315.

Mountain BW & Wood SA (1987) Solubility and transport of platinum-group elements in hydrothermal solution: Thermodynamic and physical chemical constraints. In: Prichard HM, Bowles JFW, & Cribb SJ ed. Geo-Platinum 87, Milton Keynes, United Kingdom, 22–23 April 1987. London, Elsevier Applied Science, pp 57–82.

Mountain BW & Wood SA (1988) Chemical controls on the solubility, transport, and deposition of platinum and palladium in hydrothermal solution: A thermodynamic approach. Econ Geol, 83: 492–510.

Munro-Ashman D, Munro D, & Hughes TH (1969) Contact dermatitis from palladium. Dermatol Soc, 55: 196–197.

Murdoch RD & Pepys J (1985) Cross reactivity studies with platinum group metal salts in platinum-sensitised rats. Int Arch Allergy Appl Immunol, 77: 456–458.

Murdoch RD & Pepys J (1986) Enhancement of antibody production by mercury and platinum group metal halide salts. Int Arch Allergy Appl Immunol, 80: 405–411.

Murdoch RD & Pepys J (1987) Platinum group metal sensitivity: reactivity to platinum group metal salts in platinum halide salt-sensitive workers. Ann Allergy, 59: 464–469.

Murdoch RD, Pepys J, & Hughes EG (1986) IgE antibody responses to platinum group metals: a large scale refinery survey. Br J Ind Med, 43: 37–43.

Namikoshi T, Yoshimatsu T, Suga K, Fujii H, & Yasuda K (1990) The prevalence of sensitivity to constituents of dental alloys. J Oral Rehabil, 17: 377–381.

NAS (1977) Medical and biologic effects of environmental pollutants. Platinum-group metals. Washington, DC, National Research Council, National Academy of Sciences, Division of Medical Sciences, Assembly of Life Sciences, pp 79–223.

Navarro-Ranninger C, Zamora F, Peréz JM, López-Solera I, Martínez-Carrera S, Masaguer JR, & Alonso C (1992) Palladium(II) salt and complexes of spermidine with a six-member chelate ring.

Synthesis, characterization, and initial DNA-binding and antitumor studies. J Inorg Biochem, **46**: 267–279.

Navarro-Ranninger C, Pérez JM, Zamora F, González VM, Masaguer JR, & Alonso C (1993) Palladium (II) compounds of putrescine and spermine. Synthesis, characterization, and DNA-binding and antitumor properties. J Inorg Biochem, **52**: 37–49.

Neumüller OA (1985) [Palladium.] In: Römpps Chemie-Lexikon, 8th ed. Bd. 4. Stuttgart, Franckh'sche Verlagshandlung, pp 2971–2973 (in German).

Nielson KB, Atkin CL, & Winge DR (1985) Distinct metal-binding configurations in metallothionein. J Biol Chem, **260**: 5342–5350.

Niemi L & Hensten-Pettersen A (1985) *In vitro* cytotoxicity of Ag–Pd–Cu-based casting alloys. J Biomed Mater Res, **19**: 549–561.

Nordlind K (1986) Further studies on the ability of different metal salts to influence the DNA synthesis of human lymphoid cells. Int Arch Allergy Appl Immunol, **79**: 83–85.

Nordlind K & Liden S (1993) *In vitro* lymphocyte reactivity to heavy metal salts in the diagnosis of oral mucosal hypersensitivity to amalgam restorations. Br J Dermatol, **128**: 38–41.

OECD (1992) Skin sensitisation. In: OECD guidelines for testing of chemicals — Section 4: Health effects (No. 406). Paris, Organisation for Economic Co-operation and Development, pp 1–9.

Olden K (1997) NTP chemical repository (Radian Corporation, 29 August 1991) palladium chloride. National Toxicology Program (Internet communication of 5 June 1997 at web site http://ntp-db.niehs.nih.gov/NTP_R...H&S/NTP_Chem7/Radian7647-10-1.txt).

Orestano G (1933) The pharmacologic action of palladium chloride. Boll Soc Ital Biol Sper, **8**: 1154–1156.

Panova AI & Veselov VG (1978) [Toxicity of chlorpalladosamines through chronic inhalation by experimental animals.] Gig Tr Prof Zabol, **22**: 45–46 (in Russian).

Paul AK, Mansuri-Torshizi H, Srivastava TS, Chavan SJ, & Chitnis MP (1993) Some potential antitumor 2,2'-dipyridylamine Pt(II)/Pd(II) complexes with amino acids: their synthesis, spectroscopy, DNA binding, and cytotoxic studies. J Inorg Biochem, **50**: 9–20.

Pfeiffer P & Schwickerath H (1995) [Comparison of the solubility of NEM- and palladium alloys.] Dtsch Zahnaerztl Z, **50**: 136–140 (in German).

Phielepeit T, Legrum W, Netter KJ, & Klötzer WT (1989) Different effects of intraperitoneally and orally administered palladium chloride on the hepatic monooxygenase system of male mice. Arch Toxicol, Suppl **13**: 357–362.

Pillai CKS & Nandi US (1977) Interaction of palladium (II) with DNA. Biochim Biophys Acta, **474**: 11–16.

Pistoor FHM, Kapsenberg ML, Bos JD, Meinardi MMHM, von Blomberg BME, & Scheper RJ (1995) Cross-reactivity of human nickel-reactive T-lymphocyte clones with copper and palladium. J Invest Dermatol, **105**: 92–95.

Pneumatikakis G, Yannopoulos A, & Markopoulos I (1989) Mono-vitamin B6 complexes of palladium (II) and their interactions with nucleosides. J Inorg Biochem, 37: 17–28.

Purt R (1991) [Possible health danger from palladium-based fired alloys. Quantitative determination of palladium and gallium in atmospheric zone of dental technicians.] Quintessenz Zahntech, 17: 329–334 (in German).

Puthraya KH, Srivastava TS, Amonkar AJ, Adwankar MK, & Chitnis MP (1985) Some mixed-ligand palladium (II) complexes of 2,2'-bipyridine and amino acids as potential anticancer agents. J Inorg Biochem, 25: 207–215.

Puthraya KH, Srivastava TS, Amonkar AJ, Adwankar MK, & Chitnis MP (1986) Some potential anticancer palladium (II) complexes of 2,2'-bipyridine and amino acids. J Inorg Biochem, 26: 45–54.

Rapaka RS, Sorensen KR, Lee SD, & Bhatnagar RS (1976) Inhibition of hydroxyproline synthesis by palladium ions. Biochim Biophys Acta, 429: 63–71.

Rau T & van Eldik R (1996) Mechanistic insight from kinetic studies on the interaction of model palladium(II) complexes with nucleic acid components. Met Ions Biol Syst, 32: 339–378.

Rebandel P & Rudzki E (1990) Allergy to palladium. Contact Dermatitis, 23: 121–122.

Rencz AN & Hall GEM (1992) Platinum group elements and Au in arctic vegetation growing on gossans, Keewatin District, Canada. J Geochem Explor, 43: 265–279.

Renner H (1992) Platinum group metals and compounds. In: Elvers B, Hawkins S, & Schulz G ed. Ullmann's encyclopedia of industrial chemistry, 5th ed. Weinheim, VCH Verlagsgesellschaft, pp 75–131.

Renner H & Schmuckler G (1991) Platinum-group metals. In: Merian E ed. Metals and their compounds in the environment. Weinheim, VCH Verlagsgesellschaft, pp 1135–1151.

Renner H & Tröbs U (1986) Edelmetalle. In: Harnisch H, Steiner R, & Winnacker K ed. Chemische Technologie, Band 4, Metalle, 4th ed. Munich, Carl Hanser Verlag, pp 540–572.

Reuling N ed. (1992) [Biocompatibility of dental alloys: Toxicologic, histopathologic and analytical aspects.] Munich, Wien, Carl Hanser Verlag, 163 pp (in German).

Reuling N, Fuhrmann R, & Keil M (1992) [Subacute systemic toxicity of noble metal alloys.] Dtsch Zahnaerztl Z, 47(11): 747–752 (in German).

Richter G (1996) [Dental materials — problem compounds in the allergologic diagnostic? Part II: Patch test diagnostic and relevance and evaluation of chosen dental material groups.] Hautarzt, 47: 844–849 (in German).

Richter G & Geier J (1996) [Dental materials — problem compounds in allergologic diagnostic? Part I: Analysis of the results in patients having problems with oral mucosa due to materials.] Hautarzt, 47: 839–843 (in German).

Ridgway LP & Karnofsky DA (1952) The effects of metals on the chick embryo: Toxicity and production of abnormalities in development. Ann N Y Acad Sci, 55: 203–215.

Rocklin RD (1984) Determination of gold, palladium, and platinum at the parts-per-billion level by ion chromatography. Anal Chem, **56**: 1959–1962.

Roshchin AV, Veselov VG, & Panova AI (1984) Industrial toxicology of metals of the platinum group. J Hyg Epidemiol Microbiol Immunol, **28**: 17–24.

Rosner G, Artelt S, Mangelsdorf I, & Merget R (1998) [Platinum from automotive catalytic converters: Environmental health evaluation based on recent data on exposure and effects.] Umweltmed Forsch Prax, **3**: 365–375 (in German).

Rudzki E & Prystupa K (1994) Sensitivity to various nickel and chromium concentrations in patch tests and oral challenge tests. Contact Dermatitis, **30**: 254–255.

Sabat M (1996) Ternary metal ion–nucleic acid base–protein complexes. Met Ions Biol Syst, **32**: 521–555.

Samara C & Kouimtzis TA (1987) [Preconcentration of silver(I), gold(III) and palladium (II) in water samples with 2,2'-dipyridyl-3-[(4-amino-5-mercapto)-1,2,4-triazolyl] hydrazone supported on silica gel.] Fresenius Z Anal Chem, **327**: 509–512 (in German).

Santucci B, Cristaudo A, Cannistraci C, & Picardo M (1995) Interaction of palladium ions with the skin. Exp Dermatol, **4**: 207–210.

Santucci B, Cannistraci C, Christaudo A, & Picardo M (1996) Multiple sensitivities to transition metals: the nickel palladium reactions. Contact Dermatitis, **35**: 283–286.

Sarwar M, Thibert RJ, & Benedict WG (1970) Effect of palladium chloride on the growth of *Poa pratensis*. Can J Plant Sci, **50**: 91–96.

Sax NI ed. (1979) Dangerous properties of industrial materials, 6th ed. New York, Van Nostrand Reinhold Company, pp 2113, 2272, 2279.

Sax NI & Lewis RJ ed. (1987) Hawley's condensed chemical dictionary. New York, Van Nostrand Reinhold Company, pp 869–870.

Schäfer J, Puchelt H, & Eckhardt J-D (1996) Traffic-related noble metal emissions in south-west Germany. J Conf Abstr, **1**(1): 536.

Schäfer J, Hannker D, Eckhardt J-D, & Stüben D (1998) Uptake of traffic-related heavy metals and platinum group elements (PGE) by plants. Sci Total Environ, **215**: 59–67.

Schaffran RM, Storrs FJ, & Schalock P (1999) Prevalence of gold sensitivity in asymptomatic individuals with gold dental restorations. Am J Contact Dermatitis, **10**(4): 201–206.

Schaller K-H, Angerer J, & Lehnert G (1994) [Biomonitoring in occupational and environmental medicine.] Klin Labor, **40**: 67–72 (in German).

Schedle A, Samorapoompichit P, Rausch-Fan XH, Franz A, Füreder W, Sperr WR, Sperr W, Ellinger A, Slavicek R, Boltz-Nitulescu G, & Valent P (1995) Response of L-929 fibroblasts, human gingival fibroblasts, and human tissue mast cells to various metal cations. J Dent Res, **74**: 1513–1520.

Scheff P, Wadden R, Ticho KK, Nakonechniy J, Prodanchuk M, & Hryhorczuk D (1997) Toxic air pollutants in Chernivtsi, Ukraine. Environ Int, **23**(3): 273–290.

Scheuer B, Rüther T, von Bülow V, Henseler T, Deing B, Denzer-Fürst S, Dreismann G, Eckstein L, Eggers B, Engelke H, Gärtner K, Hansen G, Hardung H, Heidbreder G, Hoffmann E, Kitzmann H, Kröger J, Mehnert I, Möhlenbeck F, Schlaak HE, Schmoll A, Schmoll M, Schröder I, Sipkova S, Sterry G, Trettel W, & Walsdorfer U (1992) [Frequently occurring contact allergens.] Aktuel Dermatol, **18**: 44–49 (in German).

Schlemmer G & Welz B (1986) Palladium and magnesium nitrates, a more universal modifier for graphite furnace atomic absorption spectrometry. Spectrochim Acta B, **41**: 1157–1165.

Schmalz G, Arenholt-Bindslev D, Pfüller S, & Schweikl H (1997a) Cytotoxicity of metal cations used in dental cast alloys. ATLA — Altern Lab Anim, **25**: 323–330.

Schmalz G, Arenholt-Bindslev D, Hiller KA, & Schweikl H (1997b) Epithelium-fibroblast co-culture for assessing mucosal irritancy of metals used in dentistry. Eur J Oral Sci, **105**: 86–91.

Schmalz G, Schuster U, & Schweikl H (1998) Influence of metals on IL-6 release *in vitro*. Biomaterials, **19**: 1689–1694.

Schnuch A & Geier J (1995) [The most common contact allergens during 1994: Data from clinics participating in the IVDK in cooperation with the German Contact Allergy Group.] Dermatosen, **43**(6): 275–278 (in German).

Schnuch A, Uter W, Lehmacher W, Fuchs T, Enders F, Arnold R, Bahmer F, Brasch J, Diepgen TL, Frosch PJ, Henseler T, Muller ST, Peters K-P, Schulze-Dirks A, Stary A, & Zimmermann J (1993) [Epicutaneous testing with standard series — first data of the project of the Information Association of Dermatological University Departments (IVDK).] Dermatosen, **41**: 60–70 (in German).

Schramel P, Wendler I, & Angerer J (1997) The determination of metals (antimony, bismuth, lead, cadmium, mercury, palladium, platinum, tellurium, thallium, tin and tungsten) in urine samples by inductively coupled plasma–mass spectrometry. Int Arch Occup Environ Health, **69**: 219–223.

Schroeder HA & Mitchener M (1971) Scandium, chromium (VI), gallium, yttrium, rhodium, palladium, indium in mice: Effects on growth and life span. J Nutr, **101**: 1431–1438.

Schultz I, Melle B, & Lenz E (1997) [Relationship between biocorrosion and noble metal content of dental alloys.] Dtsch Zahnaerztl Z, **52**: 355–360 (in German).

Schuppe HC, Rönnau AC, von Schmiedeberg S, Ruzicka T, Gleichmann E, & Griem P (1998) Immunomodulation by heavy metal compounds. Clin Dermatol, **16**: 149–157.

Schuster M & Schwarzer M (1996) Selective determination of palladium by on-line column preconcentration and graphite furnace atomic absorption spectrometry. Anal Chim Acta, **328**: 1–11.

Schwickerath H (1989) [Properties and behaviour of burning-on palladium- and non-noble metal (NEM)-alloys.] Phillip J Restaur Zahnmed, **6**: 357–367 (in German).

Shah M, Lewis FM, & Gawkrodger DJ (1997) Patch testing in children and adolescents: Five years' experience and follow-up. J Am Acad Dermatol, **37**(6): 964–968.

Shah NK & Wai CM (1985) Extraction of palladium from natural samples with bismuth diethyldithiocarbamate for neutron activation analysis. J Radioanal Nucl Chem, **94**: 129–138.

Sharkey J, Chovnick SD, Behar RJ, Perez R, Otheguy J, Solc Z, Huff W, & Cantor A (1998) Outpatient ultrasound-guided palladium 103, brachytherapy for localized adenocarcinoma of the prostate: a preliminary report of 434 patients. Urology, 51(5): 796–803.

Sheard C (1955) Contact dermatitis from platinum and related metals. Arch Dermatol Syphilol, 71: 357–360.

Shishniashvili DM, Lystsov VN, Ulanov BP, & Moshkovskii YS (1971) Investigation of the interaction of DNA with palladium ions. Biofizika, 16: 965–969.

Smith IC, Carson BL, & Ferguson TL (1978) Palladium. In: Smith IC, Carson BL, & Ferguson TL ed. Trace metals in the environment. Vol. 4. Palladium and osmium. Ann Arbor, Michigan, Ann Arbor Science, 140 pp.

Somers E (1959) Plant pathology. Fungitoxicity of metal ions. Nature, 184: 475–476.

Spikes JD & Hodgson CF (1969) Enzyme inhibition by palladium chloride. Biochem Biophys Res Commun, 35: 420–422.

Stejskal VDM, Cederbrant K, Lindvall A, & Forsbeck M (1994) MELISA — an *in vitro* tool for the study of metal allergy. Toxic In Vitro, 8: 991–1000.

Stenman E & Bergman M (1989) Hypersensitivity reactions to dental materials in a referred group of patients. Scand J Dent Res, 97: 76–83.

Stetzenbach KJ, Amano M, Kreamer DK, & Hodge VF (1994) Testing the limits of ICP-MS: Determination of trace elements in ground water at the part-per-trillion level. Ground Water, 32: 976–985.

Strietzel R & Viohl J (1992) [The long-term corrosion behaviour of NEM-, palladium alloys and titanium in artificial saliva.] Dtsch Zahnaerztl Z, 47: 535–538 (in German).

Stümke M (1992) Dental materials; metallic materials for prostheses. In: Elvers B, Hawkins S, & Schulz G ed. Ullmann's encyclopedia of industrial chemistry, 5th ed. Weinheim, VCH Verlagsgesellschaft, pp 260–264.

Suggs JW, Higgins JD, Wagner RW, & Millard JT (1989) Base-selective DNA cleavage with a cyclometalated palladium complex. In: Tullius TD ed. Metal–DNA chemistry. 195th National Meeting of the American Chemical Society, Toronto, Ontario, 5–11 June 1988. Washington, DC, American Chemical Society, pp 146–158 (ACS Symposium Series 402).

Suraikina TI, Zakharova IA, Mashkovskii Y, & Fonshtein LM (1979) Study of the mutagenic action of platinum and palladium compounds on bacteria. Cytol Genet (USSR), 13: 50–54.

Tachibana S, Murakami T, & Sawada S (1972) Studies on CO_2-fixing fermentation. (XXIII) Effects of trace elements on L-malate fermentation using *Schizophyllum commune*. J Ferment Technol, 50: 171–177.

Taubler JH (1977) Allergic response to platinum and palladium complexes — determination of no-effect level. Research Triangle Park, North Carolina, US Environmental Protection Agency (EPA-600/1-77-039).

Tayim HA, Malakian AH, & Bikhazi AB (1974) Synthesis and physicochemical, antimitogenic, and antiviral properties of a novel palladium (II) coordination compound. J Pharm Sci, 63: 1469–1471.

Taylor A, Branch S, Halls D, Owen LMW, & White MA (1998) Atomic spectrometry update; clinical and biological materials, food and beverages. J Anal Atom Spectrom, **13**: 57R–106R.

Taylor RT (1976) Comparative methylation chemistry of platinum, palladium, lead and manganese. Research Triangle Park, North Carolina, US Environmental Protection Agency (EPA-600/1-76-016).

Teicher BA, Varshney A, Khandekar V, & Herman TS (1991) Effect of hypoxia and acidosis on the cytotoxicity of six metal(ligand)$_4$(rhodamine-123)$_2$ complexes at normal and hyperthermic temperatures. Int J Hyperther, **7**: 857–868.

Tibbling L, Thuomas K-A, Lenkei R, & Stejskal V (1995) Immunological and brain MRI changes in patients with suspected metal intoxication. Int J Occup Med Toxicol, **4**: 285–294.

Tillery JB & Johnson DE (1975) Determination of platinum, palladium, and lead in biological samples by atomic absorption spectrophotometry. Environ Health Perspect, **12**: 19–26.

Todd DJ & Burrows D (1992) Patch testing with pure palladium metal in patients with sensitivity to palladium chloride. Contact Dermatitis, **26**: 327–331.

Tomilets VA & Zakharova IA (1979) [Anaphylactic and anaphylactoid properties of complex palladium compounds.] Farmakol Toksikol, **42**: 170–173 (in Russian).

Tomilets VA, Dontsov VI, Zakharova IA, & Klevtsov AV (1980) Histamine releasing and histamine binding action of platinum and palladium compounds. Arch Immunol Ther Exp (Warsz), **28**: 953–957.

Tong SSC, Morse RA, Bache CA, & Lisk DJ (1975) Elemental analysis of honey as an indicator of pollution. Arch Environ Health, **30**: 329–332.

Tripkovic M, Todorovic M, & Holclajtner-Antunovic I (1994) Spectrometric determination of gold, platinum and palladium in geological materials by d.c. arc plasma. Anal Chim Acta, **296**: 315–323.

Uno Y & Morita M (1993) Mutagenic activity of some platinum and palladium complexes. Mutat Res, **298**: 269–275.

US FDA (1973) Method of testing primary irritant substances § 191.11. In: Code of federal regulations (as special edition of the Federal Register): Title 21, Food and drugs. Washington, DC, US Government Printing Office, US Office of the Federal Register, pp 19–20.

Uter W, Fuchs T, Häusser M, & Ippen H (1995) Patch test results with serial dilutions of nickel sulfate (with and without detergent), palladium chloride, and nickel and palladium metal plates. Contact Dermatitis, **32**: 135–142.

van Joost T & Roesyanto-Mahadi ID (1990) Combined sensitization to palladium and nickel. Contact Dermatitis, **22**: 227–228.

van Ketel WG & Niebber C (1981) Allergy to palladium in dental alloys. Contact Dermatitis, **7**: 331–357.

van Loon LAJ, van Elsas PW, van Joost T, & Davidson CL (1984) Contact stomatitis and dermatitis to nickel and palladium. Contact Dermatitis, **11**: 294–297.

van Loon LAJ, van Elsas PW, van Joost T, & Davidson CL (1986) Test battery for metal allergy in dentistry. Contact Dermatitis, **14**: 158–161.

van Loon LAJ, van Elsas PW, Bos JD, ten Harkel-Hagenaar HC, Krieg SR, & Davidson CL (1988) T-lymphocyte and Langerhans cell distribution in normal and allergically induced oral mucosa in contact with nickel-containing dental alloys. J Oral Pathol, **17**: 129–137.

Vilaplana J, Romaguera C, & Cornellana F (1994) Contact dermatitis and adverse oral mucous membrane reactions related to the use of dental prostheses. Contact Dermatitis, **30**: 80–84.

Vincenzi C, Tosti A, Guerra L, Kokelj G, Nobile C, Rivara G, & Zangrando E (1995) Contact dermatitis to palladium: A study of 2300 patients. Am J Contact Dermatitis, **6**: 110–112.

Vitsentzos SJ, Vlahogiannis E, Glaros D, & Vlahomitros J (1988) The effect of fixed partial dentures made of silver–palladium alloy on serum immunoglobulins IgA, IgG and IgM. J Prosthet Dent, **59**: 587–589.

Wahlberg JE & Boman A (1990) Palladium chloride — A potent sensitizer in the guinea pig. Am J Contact Dermatitis, **1**: 112–113.

Wahlberg JE & Boman A (1992) Cross-reactivity to palladium and nickel studied in the guinea pig. Acta Derm Venereol, **72**: 95–97.

Wahlberg JE & Liden C (1999) Cross-reactivity patterns of palladium and nickel studied by repeated open applications (ROATs) to the skin of guinea pigs. Contact Dermatitis, **41**(3): 145–149.

Warocquier-Clerout R, Hachom-Nitcheu GC, & Sigot-Luizard MF (1995) Reliability of human fresh and frozen gingiva explant culture in assessing dental materials cytocompatibility. Cells Mater, **5**: 1–14.

Wataha JC & Luckwood PE (1998) Release of elements from dental casting alloys into cell-culture medium over 10 months. Dent Mater, **14**: 158–163.

Wataha JC, Craig RG, & Hanks CT (1991a) The release of elements of dental casting alloys into cell-culture medium. J Dent Res, **70**: 1014–1018.

Wataha JC, Hanks CT, & Craig RG (1991b) The *in vitro* effects of metal cations on eukaryotic cell metabolism. J Biomed Mater Res, **25**: 1133–1149.

Wataha JC, Craig RG, & Hanks CT (1992) The effects of cleaning on the kinetics of *in vitro* metal release from dental casting alloys. J Dent Res, **71**: 1417–1422.

Wataha JC, Nakajima H, Hanks CT, & Okabe T (1994a) Correlation of cytotoxicity with elemental release from mercury- and gallium-based dental alloys *in vitro*. Dent Mater, **10**: 298–303.

Wataha JC, Hanks CT, & Craig RG (1994b) *In vitro* effects of metal ions on cellular metabolism and the correlation between these effects and the uptake of the ions. J Biomed Mater Res, **28**: 427–433.

Wataha JC, Malcolm CT, & Hanks CT (1995a) Correlation between cytotoxicity and the elements released by dental casting alloys. Int J Prosthodont, **8**: 9–14.

Wataha JC, Hanks CT, & Sun Z (1995b) *In vitro* reaction of macrophages to metal ions from dental biomaterials. Dent Mater, **11**: 239–245.

Wataha JC, Luckwood PE, Frazier KB, & Khajotia SS (1999) Effect of toothbrushing on elemental release from dental casting alloys. J Prosthodont, **8**(4): 245–251.

Wedepohl KH (1995) The composition of the continental crust. Geochim Cosmochim Acta, **59**: 1217–1232.

White J & Munns DJ (1951) Inhibitory effect of common elements towards yeast growth. J Inst Brew, **57**: 175–179.

Whyte JNC & Boutillier JA (1991) Concentrations of inorganic elements and fatty acids in geographic populations of the spot prawn *Pandalus platyceros*. Can J Fish Aquat Sci, **48**: 382–390.

Wiester MJ (1975) Cardiovascular actions of palladium compounds in the unanesthetized rat. Environ Health Perspect, **12**: 41–44.

Williams MW, Hoeschele JD, Turner JE, Jacobson KB, Christie NT, Paton CL, Smith LH, Witschi HR, & Lee EH (1982) Chemical softness and acute metal toxicity in mice and *Drosophila*. Toxicol Appl Pharmacol, **63**: 461–469.

Wirz J, Jäger K, & Schmidli F (1993) [Alloy analysis (Splitter test).] In: Wirz J, Jäger K, & Schmidli F ed. Klinische Material- und Werkstoffkunde. Berlin, Quintessenz-Verlag, pp 67–78 (in German).

Wood JM, Cheh A, Dizikes LJ, Ridley WP, Rakow S, & Lakowicz JR (1978) Mechanisms for the biomethylation of metals and metalloids. Fed Proc, **37**: 16–21.

Wood SA (1991) Experimental determination of the hydrolysis constants of Pt^{2+} and Pd^{2+} at 25°C from the solubility of Pt and Pd in aqueous hydroxide solutions. Geochim Cosmochim Acta, **55**: 1759–1767.

Yang JS (1989) The comparative chemistries of platinum group metals and their periodic neighbors in marine macrophytes. In: Heavy metals in the environment: international conference. Vol. 2. Geneva, 12–15 September 1989. Edinburgh, CEP Consultants, pp 1–4.

Yoshida S, Sakamoto H, Mikami H, Onuma K, Shoji T, Nakagawa H, Hasegawa H, & Amayasu H (1999) Palladium allergy exacerbating bronchial asthma. J Allergy Clin Immunol, **103**(6): 1211–1212.

Ysart G, Miller P, Crews H, Robb P, Baxter M, De L'Argy C, Lofthouse S, Sargent C, & Harrison N (1999) Dietary exposure estimates of 30 elements from the UK total diet study. Food Addit Contam, **16**(9): 391–403.

Zereini F (1996) [Analysis of the platinum group elements (PGE) and their geochemical distribution processes in chosen sedimentary rocks and anthropogenic influenced environmental compartments in West Germany.] Frankfurt/Main, Johann Wolfgang Goethe University (professorial dissertation) (in German).

Zereini F, Zientek C, & Urban H (1993) [Concentration and distribution of platinum-group elements (PGE) in soils — Platinum emission by attrition of automobile exhaust catalysts.] Umweltwiss Schadstoff-Forsch, **5**: 130–134 (in German).

Zereini F, Alt F, Rankenburg K, Beyer J-M, & Artelt S (1997) [The distribution of the platinum group elements (PGE) in the environmental compartments of soil, mud, roadside dust, road sweepings and water — Emission of the platinum group elements (PGE) from motor vehicle catalytic converters.] Umweltwiss Schadstoff-Forsch, 9: 193–200 (in German).

Zhang Z-Q, Liu H, Zhang H, & Li Y-F (1996) Simultaneous cathodic stripping voltammetric determination of mercury, cobalt, nickel and palladium by mixed binder carbon paste electrode containing dimethylglyoxime. Anal Chim Acta, 333: 119–124.

Zhou H & Liu J (1997) The simultaneous determination of 15 toxic elements in foods by ICP-MS. Atom Spectrosc, 18: 115–118.

Zinke T (1992) [Palladium alloys.] Bundesgesundheitsblatt, 11: 579–581 (in German).

Zocchi A, Roella V, & Calamari D (1996) [Air quality assessment in the Varese area using epiphytic lichens.] Ing Ambientale, 25: 78–87 (in Italian).

RESUME

1. Identité, propriétés physiques et chimiques et méthodes d'analyse

Le palladium est un métal ductile, de couleur blanc-acier, qui ressemble aux autres métaux de la mine de platine et au nickel et qui se retrouve dans leurs minerais. Il existe sous trois degrés d'oxydation : 0 (métal), +2 et +4. Il peut former des composés organométalliques, dont seuls quelques-uns ont un usage industriel. Le palladium métallique est stable dans l'air et il resiste à l'attaque de la plupart des réactifs, sauf l'eau régale et l'acide nitrique.

Aucune méthode de dosage n'a été publiée permettant de distinguer les différentes espèces chimiques solubles ou insolubles présentes dans l'environnement.

Les méthodes couramment utilisées pour le dosage du palladium sont la spectrométrie d'absorption atomique en four à électrodes de graphite et la la spectrométrie de masse à source plasma à couplage inductif, cette dernière ayant en outre l'avantage de permettre le dosage simultané de plusieurs éléments.

2. Sources d'exposition humaines dans l'environnement

Le palladium se trouve à très faible concentration dans l'écorce terrestre (<1 µg/kg) aux côtés des autres métaux de la mine du platine. Pour ses utilisations industrielles, on l'obtient principalement comme sous-produit du raffinage du nickel, du platine et autres métaux de base. Le mode de séparation des autres métaux du groupe du platine dépend de la nature du minerai.

Les gisements importants sur le plan économique se trouvent en Russie, en Afrique du Sud et en Amérique du Nord. La production mondiale de palladium est estimée à environ 260 tonnes par an.

Le palladium et ses alliages sont utilisés comme catalyseurs dans l'industrie chimique et pétrochimique et surtout, dans l'industrie automobile. La demande de palladium pour la fabrication de pots catalytiques est passée de 24 tonnes en 1993 à 139 tonnes en 1998, lorsqu'on a adopté les pots riches en palladium pour un grand nombre de véhicules à essence.

Les applications du palladium en électronique et en électrotechnique concernent certains procédés de métallisation (pâte pelliculaire épaisse), ainsi que la fabrication de contacts et de relais.

Les alliages de palladium sont également très utilisés en art dentaire, par ex. pour la confection de couronnes et de bridges.

On ne possède pas de données sur les émissions de palladium dans l'atmosphère, l'hydrosphère et la lithosphère à partir de sources naturelles ou industrielles.

Les pots catalytiques pour automobiles sont des sources mobiles de palladium. Environ 60 % des véhicules à essence européens vendus en 1997 et beaucoup d'automobiles japonaises et américaines étaient équipés de pots catalytiques au palladium. On possède peu de données sur les émissions de palladium par les pots catalytiques monolithiques à trois voies au palladium/rhodium de conception moderne. Dans le cas d'un nouveau modèle de pot catalytique, le taux d'émission de palladium particulaire se situait dans une fourchette de 4 à 108 ng/km. Ces valeurs sont du même ordre que celles qui avaient été mesurées antérieurement avec d'autres autres pots catalytiques.

3. Transport, distribution et transformation dans l'environnement

La majeure partie du palladium présent dans la biosphère se trouve sous forme de métal ou d'oxydes, c'est-à-dire des formes pratiquement insolubles dans l'eau, résistantes à la plupart des réactions qui se produisent dans ce milieu (par ex. la décomposition abiotique ou l'oxydation par des radicaux hydroxyles) et qui ne se volatilisent pas dans l'atmosphère. Par analogie avec les autres métaux du groupe du platine, on considère que le palladium ne doit pas être biologiquement transformable.

Lorsque le pH et le potentiel redox ont la valeur voulue, les acides humiques ou fulviques fixent vraisemblablement le palladium présent dans l'environnement aquatique. On a trouvé du palladium dans les cendres d'un certain nombre de végétaux, ce qui donne à penser qu'il est plus mobile dans l'environnement que le platine et par conséquent plus biodisponible pour les plantes.

4. Concentrations dans l'environnement et exposition humaine

Alors qu'on possède beaucoup de données sur la concentration de métaux comme le plomb ou le nickel dans l'environnement, on est très peu renseigné sur le palladium. Lorsqu'on met en évidence la présence de palladium dans les eaux de surface, c'est généralement à des concentrations de 0,4 à 22 ng/litre (eaux douces) et de 19 à 70 pg/litre (eaux saumâtres). Dans le sol, on fait état de concentrations qui vont de <0,7 à 47 µg/kg. Les échantillons de sol en question ont été prélevés à proximité de routes à grande circulation.

Dans les boues d'égout, la concentration va de 18 à 260 µg/kg, mais dans des boues contaminées par des décharges provenant de fabriques locales de bijoux, on a mesuré une concentration de 4700 µg/kg. Dans l'eau de boisson, la concentration est habituellement nulle ou inférieure à 24 ng/litre. Les quelques données dont on dispose indiquent que du palladium peut être présent dans les tissus de petits invertébrés aquatiques et dans divers types de viande, de poisson, de pain et de végétaux.

L'exposition de la population générale au palladium provient essentiellement des alliages dentaires, des bijoux, de l'alimentation et des émissions des pots catalytiques.

Chez l'Homme, la dose journalière moyenne ingérée peut, semble t-il, atteindre 2 µg.

Par analogie avec le platine, on estime que la concentration du palladium dans l'air ambiant doit être inférieure à 110 pg/m^3 dans les zones urbaines où circulent des véhicules équipés de pots catalytiques. L'absorption de palladium par voie respiratoire est donc très faible. On a constaté une légère accumulation de palladium, liée à la densité du traffic et à la distance à la route, dans la poussière, le sol et les plantes situées en bordure de routes.

L'exposition environnementale par la voie orale est très importante et elle peut résulter du contact direct de la gencive avec l'alliage de palladium dont est constituée telle ou telle prothèse dentaire. Il peut également y avoir exposition cutanée par contact avec des bijoux contenant du palladium.

Les alliages dentaires sont la cause la plus fréquente d'exposition permanente au palladium. Le comportement de ces alliages vis-à-vis de la corrosion dans le milieu buccal peut dépendre de la présence d'autres métaux (comme le cuivre, le gallium ou l'indium) et des traitements subis par l'alliage. Les alliages de type cupro-palladium à haute teneur en cuivre sont probablement moins résistants à la corrosion que ceux dont la teneur en cuivre est faible. La libération de palladium à partir de prothèses dentaires varie sensiblement d'un individu à l'autre en fonction de l'état de sa denture, du matériau utilisé et de certaines habitudes personnelles, comme le fait de mâcher du « chewing-gum », par exemple. Les données cliniques relatives à l'exposition iatrogène sont d'un intérêt limité en raison de leurs insuffisances sur le plan méthodologique (nombre limité d'échantillons tissulaires, groupes témoins mal appariés). Il est donc difficile de se prononcer sur la valeur exacte de la dose journalière ingérée et la fourchette proposée de ≤ 1,5-15 µg de palladium par personne constitue une estimation grossière.

On possède quelques information sur les concentrations de palladium dans la population générale, les taux urinaires se situant notamment dans la fourchette 0,006-<0,3 µg/litre chez l'adulte.

L'exposition professionnelle au palladium et à ses sels concerne principalement les personnes employées au raffinage du métal et à la fabrication de pots catalytiques. Peu de mesures ont été faites et les résultats vont de 0,4 à 11,6 µg/m^3 en moyenne pondérée par rapport au temps sur 8 h. On ne dispose d'aucune donnée récente sur la surveillance biologique des ouvriers exposés au palladium ou à ses sels.

Les techniciens dentaires peuvent être exposés à des volumes importants de poussière de palladium lorsqu'ils travaillent ou polissent des prothèses dentaires en alliages de palladium, surtout en l'absence de dispositifs de protection (extraction ou aspiration de la poussière).

5. Cinétique et métabolisme chez l'Homme et les animaux de laboratoire

On ne possède que peu de données sur la cinétique du palladium à l'état métallique ou ionique.

Après administration à des rats par voie orale, le chlorure de palladium (II) ($PdCl_2$) a été médiocrement résorbé dans les voies digestives (moins de 0,5 % de la dose initiale chez le rat adulte ou environ 5 % chez le raton à la mamelle au bout de 3 à 4 jours). Après administration à des rats par voie intratrachéenne ou intraveineuse, la résorption/rétention était plus élevée, avec des valeurs respectivement égales à 5 et 20 % au bout de 40 jours). On a observé une résorption cutanée, mais sans la mesurer.

Après avoir administré divers dérivés du palladium par voie intraveineuse à des rats, des lapins et des chiens, on a décelé la présence de cet élément dans plusieurs tissus de ces animaux. C'est dans le rein, le foie, la rate, les ganglions lymphatiques, les surrénales, le poumon et les os que l'on a retrouvé les concentrations les plus élevées. Par exemple, 8 à 21 % de la dose de chlorure de palladium (II) ou de tétrachloropalladate (II) de sodium (Na_2PdCl_4) ont été retrouvés dans le foie ou les reins de rats 1 jour après l'administration. Après avoir fait ingérer du monoxyde de palladium (II) (PdO) à des rats pendant 4 semaines avec leur nourriture, on n'en a retrouvé des concentrations mesurables que dans les reins.

On ne dispose que de maigres informations sur la distribution tissulaire ou liquidienne du palladium (par ex. dans le sérum et la salive, environ 1 µg/litre) utilisé pour des interventions de réfection dentaire.

Après administration d'une dose unique de chlorure de palladium (II) par voie intraveineuse à des rats, on a observé le passage de faibles quantités de palladium dans la progéniture par l'intermédiaire du lait ou par voie placentaire.

On est peu renseigné sur l'élimination et l'excrétion du palladium et ce que l'on en sait concerne essentiellement le chlorure de palladium (II) et le tétrachloropalladate (II) de sodium, que l'on retrouve dans les matières fécales et les urines. Chez des rats et des lapins ayant reçu ces composés par voie intraveineuse pendant 3 h à 7 jours, on a observé des taux d'excrétion urinaire allant de 6,4 à 76 % de la dose initiale. Ces études ont également montré que que l'élimination du palladium par la voie fécale allait de traces à 13 % de la dose administrée. Après administration de chlorure de palladium (II) par voie orale à des rats, plus de 95 % du palladium a été éliminé dans les matières fécales par suite de sa non résorption. L'application topique ou l'injection sous-cutanée de sulfate de palladium (II) ($PdSO_4$) ou d'autres dérivés du palladium à des cobayes et à des lapins s'est traduite par la présence de palladium à un niveau décelable dans les urines.

Le calcul de la demi-vie d'élimination du palladium chez le rat (corps entier, foie, rein) donne un chiffre compris entre 5 et 12 jours.

La mesure du taux de rétention à intervalles de 3 h, 24 h et 48 h après injection intraveineuse de $^{103}PdCl_2$ à des rats, n'a guère varié au cours du temps en ce qui concerne le rein, la rate, les muscles, le pancréas, le thymus, le cerveau et les os. On a noté une diminution légère dans le foie et sensible dans le poumon, les surrénales et le sang.

En raison de la facilité avec laquelle les ions palladium forment des complexes, ils se lient aux acides aminés comme la L-cystéine, la L-cystine ou la L-méthionine, aux protéines (par ex. la caséine, la fibroïne de la soie et de nombreuses enzymes), à l'ADN et à d'autres macromolécules, comme la vitamine B_6.

De nombreuses études confirment l'affinité des dérivés du palladium pour les acides nucléiques. Des tests *in vitro* portant sur le chlorure de palladium (II) en présence de thymus de veau montrent que le palladium (II) réagit à la fois sur les groupements phosphate et les bases de l'ADN. On a constaté que plusieurs complexes organopalladiens formaient des liaisons avec l'ADN du thymus de veau et l'ADN plasmidique d'*Escherichia coli*. Il semble que la plupart du temps, les liaisons formées par ces complexes soient non covalentes et qu'il s'agisse principalement de liaisons hydrogène. Cependant il existe quelques cas où les liaisons formées sont covalentes.

6. **Effets sur les mammifères de laboratoire et les systèmes d'épreuve *in vitro***

Selon la nature du composé et la voie d'exposition étudiée, la DL_{50} pour le palladium varie de 3 à >4900 mg/kg, le composé le plus toxique étant le chlorure de palladium (II) et le moins toxique, l'oxyde de palladium (II). C'est par la voie orale que la toxicité est la plus faible. Dans le cas de la voie intraveineuse, la DL_{50} est du même ordre pour les composés suivants : chlorure de palladium (II), tétrachloro-palladate (II) de potassium (K_2PdCl_4) et tétrachloropalladate (II) d'ammonium (($NH_4)_2PdCl_4$). Dans le cas du chlorure de palladium (II) on a pu mettre en évidence des différences marquées tenant à la voie d'administration : par exemple, avec des rats Charles-River CD1, la valeur de la DL_{50} est de 5 mg/kg de poids corporel pour la voie intra-veineuse, de 6 mg/kg p.c. pour la voie intratrachéenne, de 70 mg/kg p.c. pour la voie intrapéritonéale et de 200 mg/kg p.c. pour la voie orale. Avec des rats Sprague-Dawley, la DL_{50} par voie orale est plus élevée.

La toxicité aiguë de plusieurs composés du palladium administrés à des rats ou à des lapins s'est manifestée de la manière suivante : mortalité, diminution de la consommation d'eau et de nourriture, émaciation, ataxie et déplacement sur la pointe des pattes, convulsions toniques et cloniques, effets cardiovasculaires, péritonite, modification des paramètres biochmiques (par ex. modification de l'activité des enzymes hépatiques, protéinurie ou cétonurie). On a également observé des anomalies fonctionnelles et histologiques au niveau du rein, tant après administration de dérivés du palladium que du métal à l'état pulvérulent. Des hémorragies ont également mises en évidence au niveau des poumons et du grêle.

Les anomalies constatées chez les rongeurs et les lapins après une exposition de courte durée à divers dérivés du palladium concernent principalement des paramètre biochimiques (par ex. une diminution de l'activité des enzymes microsomiques hépatiques ou de la formation de protéines microsomiques). Les signes cliniques consistaient en apathie, perte de poids, hématomes ou exudations. On a également observé une modification du poids absolu et relatif des viscères. L'un des composés - le tétrachloropalladate (II) de sodium, complexé par de l'albumine d'oeuf - a provoqué la mort de souris. Les concentrations agissantes étaient de l'ordre du milligramme par kg de poids corporel. Des effets histopathologiques ont été observés au niveau du foie, du rein, de la rate ou de la muqueuse gastrique de rats, 28 jours après administration par voie orale de 15 ou 150 mg d'hydrogénocarbonate de tétrammine palladium ($[Pd(NH_3)_4](HCO_3)_2$) par kg de poids corporel. En outre, on a constaté une augmentation du poids absolu du cerveau et des ovaires aux doses de 1,5 et 15 mg par kg de poids corporel.

On ne sait pas exactement quelle est la contribution du palladium aux effets observés après administration d'un produit formé d'un alliage dentaire contenant ce métal, soit en une seule fois, soit pendant une courte période.

Par ailleurs, on ne possède guère de données sur les effets à long terme d'une exposition au palladium sous différentes formes chimiques.

Des souris auxquelles on avait administré du chlorure de palladium (II) à raison de 5 mg de palladium par litre dans leur eau de boisson depuis le sevrage jusqu'à la mort naturelle, ont présenté une dépression pondérale, un allongement de la durée de vie (uniquement chez les mâles), une amyloïdose prononcéeau niveau de plusieurs viscères et une multiplication environ par deux de la fréquence des tumeurs malignes (voir plus loin).

L'exposition par voie respiratoire de rats à de la chloropalladosammine ($(NH_3)_2PdCl_2$) pendant environ six mois, a provoqué une modification légère - réversible à 5,4 mg/m^3 et permanente à 18 mg/m^3 - de plusieurs paramètres sériques ou urinaires, ce qui est l'indication de lésions prrincipalement au niveau du foie et du rein (une réduction du gain de poids, une modification du poids des organes et une glomérulonéphrite ont également été observées). L'exposition par la voie entérique a également entraîné des effets indésirables, la dose sans effet nocif observable étant de 0,08 mg/kg de poids corporel.

Six mois après administration intratrachéenne à des rats de poussière de palladium à raison de 143 mg/kg de poids corporel, plusieurs signes histopathologiques ont été relevés au niveau du poumon. L'administration quotidienne par voie orale de ce type de poussière à des rats (à raison de 50 mg/kg de poids corporel) pendant 6 mois a entraîné la modification de plusieurs paramètres sériques et urinaires.

Des tests cutanés effectués sur des lapins à l'aide d'une série de dérivés du palladium ont permis de mettre en évidence des réactions variables, qui se rangeaient dans l'ordre de gravité suivant : $(NH_4)_2PdCl_6$ > $(NH_4)_2PdCl_4$ > $(C_3H_5PdCl)_2$ > K_2PdCl_6 > K_2PdCl_4 > $PdCl_2$ > $(NH_3)_2PdCl_2$ > PdO. Les trois premiers composés ont provoqué des érythèmes, des oedèmes et des escarres sur la peau intacte ou abrasée, les trois suivants ont également provoqué des érythèmes sur la peau abrasée et les deux dernières n'ont pas eu d'effet irritant. Du chlorhydrate de palladium (formule non indiquée) a également provoqué une dermatite chez le lapin.

On a observé une irritation oculaire avec le chlorure de palladium (II) et l'hydrogénocarbonate de tétrammine-palladium (mais pas avec l'oxyde de palladium (II) après dépôt de ces substances sur la surface de l'oeil de lapins. Une exposition à la chloropalladosammine (≥ 50 mg/m^3) a eu des effets nocifs sur la muqueuse oculaire de rats (conjonctivite, kératoconjonctivite).

On a constaté que certains dérivés du palladium se comportent comme de puissants sensibilisateurs cutanés (le chlorure de palladium (II), l'hydrogénocarbonate de palladium-tétrammine, le chlorhydrate de palladium (formule non indiquée) et les complexes palladium-albumine). Le chlorure de palladium (II) s'est révélé un sensibilisateur plus puissant que le sulfate de nickel ($NiSO_4$) lors du test de maximisation chez le cobaye. Des cobayes sensibilisés avec du chromate ou des sels de cobalt ou de nickel, n'ont pas réagi lors d'une épreuve au chlorure de palladium (II). Par contre, lorsqu'on les sensibilisait avec du chlorure de palladium (II), il réagissaient à une exposition au sulfate de nickel. Des résultats quelque peu divergents on été obtenus lors de l'étude de la réactivité croisée entre le palladium et le nickel par applications répétées sur la peau de cobayes (test d'usage de type ROAT). Au cours de ces expériences, on a sensibilisé les animaux avec du chlorure de palladium (II) ($n = 27$) ou du sulfate de nickel ($n = 30$) selon la méthode du test de maximisation sur cobaye, puis on les a traités topiquement une fois par jour pendant 10 jours selon la méthode employée pour un test d'usage de type ROAT avec un allergène sensibilisant (chlorure de palladium (II) et sulfate de nickel), avec le composé soupçonné de provoquer une sensibilisation croisée (sulfate de nickel ou chlorure de palladium (II) ou encore avec l'excipient contenant ces composés. Cette étude n'a pas permis de déterminer avec certitude si la réactivité au $PdCl_2$ des animaux sensibilisés par le $NiSO_4$

était une réactivité croisée ou si elle était due à Ja sensibilisation provoquée par les traitements répétés. Par contre, la réactivité vis-à-vis du $NiSO_4$ présentée par les animaux sensibilisés avec du chlorure de palladium (II) pouvait être considérée comme une réactivité croisée. On a observé une sensibilisation respiratoire (bronchospasmes) chez des chats ayant reçu plusieurs complexes du palladium par voie intaveineuse. Cette sensibilisation s'est accompagnée d'une augmentation du taux sérique d'histamine. On a obtenu des réponses immunitaires importantes avec du $PdCl_2$ ou des chloropalladates en utilisant le test sur les ganglions lymphatiques poplités et auriculaires de souris BALB/c. Les premières données obtenues sur un modèle animal incitent à penser que les dérivés du palladium (II) pourraient avoir une part de responsabilité dans l'apparition d'une maladie autoimmune.

On ne possède pas suffisamment de données concernant les effets que le palladium et ses dérivés pourraient avoir sur la reproduction ou le développement. Lors d'une étude de criblage, on a constaté une réduction du poids testiculaire chez des souris qui avaient reçu quotidiennement des 30 doses de chlorure de palladium (II) par voie sous-cutanée jusqu'à un total de 3,5 mg/kg de poids corporel.

Il est possible que les composés du palladium réagissent *in vitro* sur l'ADN isolé. Cependant, à une exception près, les tests de mutagénicité effectués *in vitro* avec un certain nombre de ces composés sur des cellules bactériennes ou mammaliennes (test d'Ames sur *Salmonella typhimurium*; chromotest SOS sur *Escherichia coli*; test des micronoyaux sur lymphocytes humains) ont donné des résultats négatifs.

Deux études ont mis en évidence des tumeurs attribuées à une exposition au palladium. Des souris auxquelles on avait administré du chlorure de palladium (II) dissous dans leur eau de boisson (5 mg de Pd^{2+}/litre) depuis le moment du sevrage jusqu'à la mort naturelle, présentaient des tumeurs malignes, principalement à type de lymphome ou de leucémie ou encore des adénocarcinomes pulmonaires. Ces tumeurs se sont produites avec une fréquence statistiquement significative mais elles coïncidaient également avec une longévité accrue chez les mâles, qui pourrait expliquer au moins en partie l'augmentation de la fréquence tumorale. Des tumeurs ont été observées au bout de 504 jours chez 7 rats sur 14 aux points où des fragments d'un alliage d'argent, d'or et de palladium avaient été implantés par voie sous-cutanée (on n'a pas pu déterminer avec certitude si ces tumeurs étaient dues à un stimulus physique permanent ou aux composés chimiques eux-mêmes). On ne dispose d'aucune étude de cancérogénicité utilisant l'inhalation comme voie d'exposition.

Les ions palladium sont capables d'inhiber la plupart des grandes fonctions cellulaires, comme le montrent les études *in vitro* et *in vivo*. Il semble que la cible la plus sensible de cette action inhibitrice soit la biosynthèse de l'ADN et de l'ARN. Dans une étude *in vitro* sur fibroblastes de souris, on a constaté que la CE_{50} pour l'inhibition de la synthèse de l'ADN par le chlorure de palladium (II) était de 300 µmol/litre (soit 32 mg de Pd^{2+}/litre). On a également observé une inhibition de la synthèse de l'ADN *in vivo* (dans la rate, le foie, le rein et le testicule) chez des rats à qui l'on avait administré une seule dose de 14 µmol par kg p.c. (soit 1,5 mg de Pd^{2+}/litre) de nitrate de palladium (II) ($Pd(NO_3)_2$) par voie intrapéritonéale.

En applications sous forme métallique, le palladium semble à peu près dénué de cytotoxicité *in vitro*, ainsi qu'en témoigne l'examen microscopique.

On a constaté qu'une série d'enzymes isolées possédant des fonctions métaboliques essentielles étaient inhibées par des sels simples ou complexes de palladium. L'inhibition la plus forte (valeur de K_i pour le chlorure de palladium (II) = 0,16 µmol/litre) a été observée dans le cas de la créatine-kinase, une enzyme qui joue un rôle important dans le métabolisme énergétique.

De nombreux complexes organopalladiens ont des propriétés anti-cancéreuses analogues à celles du *cis*-dichloro-2,6-diaminopyridine-platine (II), un anticancéreux appelé aussi *cis*-platine.

Le mode d'action des ions palladium et du palladium élémentaire n'est pas parfaitement élucidé. Il est probable qu'au départ, la formation d'ions palladium complexes avec des constituants cellulaires est à la base de leur action. Il se peut que des phénomènes d'oxydation dus à la présence du palladium à différents degrés d'oxydation se produisent également.

7. Effets sur l'Homme

On ne possède aucune donnée concernant les effets que le palladium émis par les pots catalytiques des automobiles exerce sur la population dans son ensemble. On a fait état d'effets dus à une exposition iatrogène ou autre.

La plupart des rapports médicaux concernent des cas de sensibilité au palladium lors d'une réfection dentaire au moyen d'alliage à base de palladium avec les symptômes suivants : dermatite de contact, stomatite ou inflammation des muqueuses et lichen plan de la muqueuse buccale. Les patients qui font une réaction positive par apposition d'un timbre imprégné de chlorure de palladium (II) ne réagissent pas forcément au palladium métallique. Seules quelques-unes des personnes présentant une réaction positive au timbre imprégné de $PdCl_2$ on présenté des symptômes clinique au niveau de la muqueuse buccale après exposition à un alliage contenant du palladium. Une étude a révélé des modifications légères et non significatives des immunoglobulines sériques après une réfection dentaire au moyen d'un alliage d'argent et de palladium.

Parmi les autres effets secondaires observés après utilisation de palladium à des fins médicales ou expérimentales, on peut citer de la fièvre, une hémolyse, une coloration anormale ou une nécrose au point d'injection après des injections sous-cutanées ainsi qu'un érythème et un oedème après application topique.

Quelques rapports médicaux font état d'anomalies cutanées chez des patients ayant porté des bijoux contenant du palladium ou exposés à une source de palladium indéterminée.

Une série de tests effectués avec des timbres imprégnés de chlorure de palladium (II) a révélé une forte sensibilité au palladium dans les groupes particuliers étudiés. Selon des études de grande envergure récemment effectuées dans différents pays, la fréquence de la sensibilité au palladium est de 7-8 % chez les patients des services de dermatologie et les écoliers, les personnes jeunes et de sexe féminin étant plus particulièrement touchées. Comparativement aux autres allergènes (environ 25 substances ont été étudiées), le palladium figure parmi les sept substances sensibilisatrices qui provoquent les réactions les plus fréquentes (parmi les métaux, il vient en seconde position, juste après le nickel). Les réactions limitées au seul palladium (mono-allergie) sont peu fréquentes. La plupart du temps, on observe des réactions à plusieurs métaux (multisensibilité) et en premier lieu, au nickel.

Jusqu'ici, ce sont les alliages utilisés pour les travaux de réfection dentaire et les bijoux qui constituent les sources les plus fréquemment mises en cause dans les cas de sensibilité au palladium au sein de la population générale.

On possède quelques données sur les effets indésirables d'une exposition professionnelle au palladium. Parmi des ouvriers travaillant sur des métaux de la mine du platine, quelques-uns (2/307; 3/22) ont présenté une réaction positive à des tests de sensibilisation effectués avec un halogénure complexe de palladium (intradermoréaction, technique RAST ou anaphylaxie cutanée passive sur le singe). Certains travailleurs (4/130) d'une usine fabricant des pots catalytiques pour automobiles ont présenté une intradermoréaction positive au chlorure de palladium (II). Une mise au point fait état, sans donner de détails, de maladies allergiques des voies respiratoires, de dermatoses et d'affections oculaires parmi des ouvriers russes travaillant sur les métaux de la mine du platine. Des cas confirmés de dermatite de contact ont été observés chez deux chimistes et un ouvrier métallurgiste. Un unique cas d'asthme professionnel dû à une exposition à des sels de palladium a été observé dans l'industrie électronique.

Les sous-groupes de population particulièrement exposés à un risque d'allergie au palladium sont les personnes qui sont déjà allergiques au nickel.

8. Effets sur les autres êtres vivants au laboratoire et dans leur milieu naturel

On a constaté que plusieurs dérivés du palladium étaient dotés d'une activité antivirale, antibactérienne ou fongicide. Les tests habituels de toxicité microbienne n'ont été que rarement pratiqués dans des conditions simulant celles de l'environnement. Dans le cas de l'hydrogénocarbonate de tétrammine-palladium, on a obtenu une CE_{50}

à 3 h de 35 mg/litre (12,25 mg de palladium par litre) pour l'inhibition de la respiration des boues d'égout activées.

Les dérivés du palladium testés sur des organismes aquatiques se sont révélés sensiblement toxiques. Deux complexes de palladium (le tétrachloropalladate (II) de potassium et la chloropalladosammine) présents dans une solution nutritive ont provoqué la nécrose de la jacinthe d'eau (*Eichhornia crassipes*) à la concentration de 2,5-10 mg de palladium par litre. La toxicité aiguë (CL_{50} à 96 h) du chlorure de palladium (II) pour le ver tubificide dulçaquicole *Tubifex tubifex* s'est révélée égale à 0,09 mg de palladium par litre. Dans le cas d'un poisson d'eau douce (*Oryzias latipes*), on a obtenu le chiffre de 7 mg/litre de chlorure de palladium (II), soit 4,2 mg de palladium par litre, pour la concentration létale minimum à 24 h. Dans tous les cas, la toxicité des dérivés du palladium est similaire à celle des dérivés du platine.

Les tests de toxicité exécutés sur des organismes aquatiques conformément recommandations de l'Organisation pour la coopération et le développement économiques (OCDE) ne portent que sur l'hydrogénocarbonate de tétrammine-palladium. Ils ont fourni les valeurs suivantes : 0,066 mg/litre (soit 0,02 mg de palladium par litre) pour la CE_{50} à 72 h (inhibition de la multiplication cellulaire sur *Scenedesmus subspicatus*); 0,22 mg/litre, soit 0,08 mg de palladium par litre pour la CE_{50} à 48 h (immobilisation de *Daphnia magna*) et 0,53 mg/litre, soit 0,19 mg de palladium par litre, pour la CL_{50} à 96 h (toxicité aiguë pour la truite arc-en-ciel, *Oncorhynchus mykiss*). Valeurs de la concentration sans effet nocif observable (NOEC) : 0,04 mg/litre (0,014 mg de palladium par litre) pour les algues, 0,10 mg/litre (0,05 mg de palladium par litre) pour *Daphnia magna* et 0,32 mg/litre (0,11 mg de palladium par litre) pour les poissons. Toutes ces valeurs sont basées sur les concentrations nominales. On a cependant constaté que les concentrations mesurées étaient beaucoup plus faibles et variables, pour des raisons qui restent indéterminées. En ce qui concerne le test d'immobilisation de la daphnie, on a calculé des valeurs en se basant sur les concentrations moyennes mesurées et pondérées en fonction du temps : on trouve alors pour la CE_{50} à 48 h une valeur de 0,13 mg/litre (soit 0,05 mg de palladium par litre) et pour la NOEC, une valeur de 0,06 mg/litre (soit 0,02 mg de palladium par litre). On a également observé des effets phytotoxiques sur des plantes terrestres après adjonction de chlorure de palladium (II) à leur solution nutritive. Ces effets étaient les suivants : une inhibition de la transpiration à la concentration de 3 mg/litre (1,8 mg de palladium par litre), des anomalies histologiques à la concentration de 10 mg/litre (6 mg de palladium par litre) et la mort de la plante à la concentration de 100 mg/litre (60 mg de palladium par litre) chez le pâturin des prés (*Poa pratensis*). Chez plusieurs plantes vivrières, on a observé un retard de croissance et un étiolement radiculaire, les plus sensibles étant l'avoine qui a souffert à la concentration d'environ 0,22 mg de chlorure de palladium (II) par litre (soit 0,132 mg de palladium par litre).

On n'a trouvé dans la littérature aucune information concernant les effets du palladium sur les vertébrés et les invertébrés terrestres.

On ne dispose d'aucune observation sur le terrain.

RESUMEN

1. Identidad, propiedades físicas y químicas y métodos analíticos

El paladio es un elemento metálico dúctil de color blanco acero semejante a otros metales del grupo del platino y al níquel, con los que se encuentra. Existe en tres estados: Pd^0 (metálico), Pd^{2+} y Pd^{4+}. Puede formar compuestos organometálicos, de los cuales sólo se han encontrado usos industriales para unos pocos. El metal de paladio es estable en el aire y resistente al ataque de la mayoría de los reactivos, salvo el agua regia y el ácido nítrico.

Hasta ahora no se han publicado métodos de medición que permitan distinguir entre las diferentes especies de paladio soluble o insoluble en el medio ambiente.

Los métodos analíticos utilizados habitualmente para la cuantificación de los compuestos de paladio son la espectrometría de absorción atómica en horno de grafito y la espectrometría de masas de plasma con acoplamiento inductivo, permitiendo este último el análisis simultáneo de elementos múltiples.

2. Fuentes de exposición humana y ambiental

El paladio se encuentra junto con otros metales del grupo del platino en concentraciones muy bajas (<1 μg/kg) en la corteza terrestre. Para usos industriales, se recupera fundamentalmente como subproducto durante la purificación del níquel, el platino y otros metales básicos. Su separación de los metales de platino depende del tipo de mineral en el que se encuentre.

Existen fuentes importantes desde el punto de vista económico en Rusia, Sudáfrica y América del Norte. La extracción de paladio en todo el mundo se estima en unas 260 toneladas/año.

El paladio y sus aleaciones se utilizan como catalizadores en la industria (petro)química y, en particular, en la del automóvil. La demanda de paladio para catalizadores de los automóviles se elevó de

24 toneladas en 1993 a 139 toneladas en 1998, tras la adopción de una tecnología con alto contenido de paladio para numerosos automóviles de gasolina.

Entre las aplicaciones en la electrónica y la tecnología eléctrica figuran el uso en los procesos de metalización (pasta de película gruesa), contactos eléctricos y sistemas de conmutación.

Las aleaciones de paladio también se utilizan ampliamente en odontología (por ejemplo, para coronas y puentes).

No se dispone de datos cuantitativos sobre las emisiones de paladio en la atmósfera, la hidrosfera o la geosfera a partir de fuentes naturales o industriales.

Los catalizadores de los automóviles son fuentes móviles de paladio. Alrededor del 60% de los automóviles europeos de gasolina vendidos en 1997, y también muchos de los vehículos del Japón y los Estados Unidos, estaban equipados con catalizadores que contenían paladio. Apenas se dispone de datos relativos a las tasas exactas de emisiones de paladio de los automóviles equipados con catalizadores monolíticos modernos de tres vías con paladio/rodio. El paladio particulado que emite un catalizador nuevo que lo contiene oscila entre 4 y 108 ng/km. Estos valores son del mismo orden de magnitud que los notificados anteriormente para las emisiones de platino de los catalizadores.

3. Transporte, distribución y transformación en el medio ambiente

La mayor parte del paladio se encuentra en la biosfera en forma de metal o de óxidos metálicos, que son prácticamente insolubles en agua, resistentes a la mayoría de las reacciones que se producen en la biosfera (por ejemplo, la degradación abiótica, la radiación ultra-violeta, la oxidación por radicales hidroxilo) y no se volatilizan en el aire. Por analogía con otros metales del grupo del platino, no cabe esperar la transformación biológica del paladio metálico.

En condiciones adecuadas de pH y potencial de oxidación-reducción, se supone que en el medio acuático se unen al paladio péptidos o los ácidos húmico o fúlvico. Se ha encontrado paladio en las cenizas de algunas plantas, lo que parece indicar que este metal es más móvil en el medio ambiente y, por consiguiente, más biodisponible que el platino para las plantas.

4. Niveles ambientales y exposición humana

A diferencia de la abundante información sobre las concentraciones de metales como el plomo o el níquel en el medio ambiente, la relativa al paladio es escasa. Las concentraciones de paladio en las aguas superficiales, cuando se detecta, generalmente oscilan entre 0,4 y 22 ng/l (agua dulce) y entre 19 y 70 pg/l (agua salada). Las concentraciones notificadas en el suelo varían entre <0,7 y 47 µg/kg. Todas estas muestras de suelo se recogieron en zonas cercanas a carreteras importantes.

Las concentraciones notificadas en los fangos cloacales son del orden de 18 a 260 µg/kg, aunque se ha notificado una concentración de 4700 µg/kg en un fango contaminado por descargas procedentes de la industria de la joyería local. Las muestras de agua de bebida no suelen contener paladio o su concentración es <24 ng/l. Los pocos datos disponibles ponen de manifiesto que puede haber paladio presente en los tejidos de pequeños invertebrados acuáticos, diferentes tipos de carne, pescado, pan y plantas.

La población general está fundamentalmente expuesta al paladio a través de las aleaciones dentales, la joyería, los alimentos y las emisiones de los catalizadores de los automóviles.

El promedio de la ingesta humana de paladio con los alimentos parece ser de hasta 2 µg/día.

Por analogía con el platino, cabe prever concentraciones de paladio inferiores a 110 pg/m^3 en el aire ambiente de zonas urbanas donde se utilizan catalizadores de paladio. Por consiguiente, la absorción de paladio por inhalación es muy baja. En muestras de polvo, de tierra y de hierba tomadas a los lados de carreteras se ha detectado una ligera acumulación de paladio, en correlación con la densidad del tráfico y la distancia de la carretera.

La exposición oral en el medio ambiente general es muy importante y puede producirse mediante el contacto directo diario de la encía con aleaciones dentales de paladio. Se puede producir exposición cutánea por contacto con joyas que contienen paladio.

Las aleaciones dentales son la causa más frecuente de exposición constante al paladio. La acción corrosiva de estas aleaciones en la boca puede modificarse por la adición de otros metales (por ejemplo cobre, galio e indio) y la elaboración de la aleación. Las aleaciones de paladio-cobre con un contenido elevado de cobre pueden ser menos resistentes a la corrosión que las aleaciones de paladio cuyo contenido en cobre es bajo. El paladio que se desprende de los arreglos dentales que contienen paladio muestra una variación individual importante en función de las condiciones dentales, el material utilizado y los hábitos personales (por ejemplo, mascar chicle). Los datos clínicos relativos a la exposición iatrogénica tienen un valor limitado, puesto que los pocos estudios de casos realizados tienen deficiencias metodológicas, por ejemplo un número limitado de muestras de tejidos y escasa concordancia con grupos testigos. Por consiguiente, es difícil cuantificar con exactitud la ingesta diaria, de manera que el valor propuesto de $\leq 1,5$-15 µg de paladio/día por persona sigue siendo una estimación elemental.

Hay alguna información sobre las concentraciones de paladio en la población general, siendo los niveles en la orina de 0,006 a <0,3 µg/l en adultos.

La mayor parte de las exposiciones ocupacionales al paladio (sales) se producen durante la purificación del paladio y la fabricación de catalizadores. Hay pocas mediciones de la exposición, que oscila entre 0,4 y 11,6 µg/m^3 como promedio ponderado por el tiempo en ocho horas. No se dispone de datos recientes para la vigilancia biológica de los trabajadores expuestos al paladio y sus sales.

Los técnicos dentales pueden estar expuestos a concentraciones máximas de polvo de paladio durante la elaboración y el pulido de aleaciones que contienen este metal, en particular si no se adoptan medidas de protección adecuadas (técnicas de extracción o aspiración del polvo).

5. Cinética y metabolismo en animales de laboratorio y en el ser humano

Son pocos los datos disponibles sobre la cinética del paladio metálico o iónico.

La absorción de cloruro de paladio (II) ($PdCl_2$) a partir del tracto digestivo fue escasa (<0,5% de la dosis oral inicial en ratas adultas o alrededor del 5% en las ratas lactantes después de tres o cuatro días). La absorción/retención en ratas adultas fue superior tras la exposición intratraqueal o intravenosa, produciéndose, 40 días después de la administración, acumulaciones corporales totales del 5% o el 20%, respectivamente, de la dosis administrada. Se observó absorción tras la aplicación cutánea, pero no se cuantificó.

Tras la administración intravenosa de diferentes compuestos de paladio se detectó su presencia en varios tejidos de ratas, conejos o perros. Las concentraciones más altas se observaron en el riñón, el hígado, el bazo, los nódulos linfáticos, las glándulas suprarrenales, el pulmón y los huesos. Por ejemplo, un día después de la administración de cloruro de paladio (II) o de tetracloropaladato (II) de sodio (Na_2PdCl_4) se encontró en el hígado o el riñón de ratas una concentración del 8% al 21% de la dosis administrada. Tras la administración de óxido de paladio (II) (PdO) con los alimentos durante cuatro semanas, sólo se encontraron concentraciones medibles en el riñón de ratas.

Los datos sobre la distribución del paladio procedente de los arreglos dentales en los tejidos o los fluidos humanos son escasos (por ejemplo, en el suero y la saliva es de alrededor de 1 µg/l).

Se observó la transferencia de pequeñas cantidades de paladio a las crías a través de la placenta y la leche tras la administración a ratas de dosis intravenosas únicas de cloruro de paladio (II).

La información sobre la eliminación y excreción de paladio es escasa y se refiere principalmente al cloruro de paladio (II) y al tetracloropaladato (II) de sodio, cuya eliminación se observó que se efectuaba en las heces y la orina. Las tasas de excreción urinaria de ratas y conejos tratados por vía intravenosa oscilaron entre el 6,4% y el 76% de la dosis administrada durante un período de tres horas a siete días. La eliminación del paladio en las heces variaba en estos estudios entre una cantidad ínfima y el 13% de la dosis administrada. Tras la administración oral de cloruro de paladio (II), >95% del paladio se eliminaba en las heces de las ratas, debido a la no absorción. La aplicación subcutánea o cutánea de sulfato de paladio (II) ($PdSO_4$) o de otros compuestos de paladio produjo concentraciones detectables de paladio en la orina de cobayas y conejos.

La semivida calculada para la eliminación del paladio de las ratas (cuerpo entero, hígado, riñón) fue de 5 a 12 días.

Los valores medios de la retención determinados para tres períodos de tiempo (3 h, 24 h, 48 h) en ratas tratadas por vía intravenosa con $^{103}PdCl_2$ mostraron pequeños cambios a lo largo del tiempo para el riñón, el bazo, el músculo, el páncreas, el timo, el cerebro y los huesos. Se observó una disminución ligera en el hígado y pronunciada en el pulmón, las glándulas suprarrenales y la sangre.

Los iones de paladio, gracias a su capacidad para formar complejos, se unen a aminoácidos (por ejemplo, la L-cisteína, la L-cistina y la L-metionina), proteínas (por ejemplo, la caseína, la fibroína de la seda y numerosas enzimas), ADN u otras macromoléculas (por ejemplo, la vitamina B_6).

Numerosos estudios confirmaron la afinidad de los compuestos de paladio por los ácidos nucleicos. En experimentos *in vitro* con cloruro de paladio (II) y ADN de timo de ternero se puso de manifiesto una interacción del paladio (II) tanto con los grupos fosfato del ADN como con sus bases. Se observó que varios complejos orgánicos de paladio se unían al ADN del timo de terneros o al de plasmidios de *Escherichia coli*. En la mayoría de los complejos la interacción parece producirse mediante una unión no covalente, principalmente a través de un enlace de hidrógeno; sin embargo, en algunos casos se han observado indicios de un enlace covalente.

6. Efectos en mamíferos de laboratorio y en sistemas de prueba *in vitro*

Los valores de la DL_{50} para los compuestos de paladio fueron, en función del compuesto y de la vía sometida a prueba, de 3 a >4900 mg/kg de peso corporal, siendo el cloruro de paladio (II) el compuesto más tóxico y el óxido de paladio (II) el menos tóxico. La toxicidad más baja se observó en la administración oral. Se obtuvieron valores muy semejantes de la DL_{50} por vía intravenosa para el cloruro de paladio (II), el tetracloropaladato (II) de potasio (K_2PdCl_4) y el tetracloropaladato (II) de amonio ((NH_4)$_2PdCl_4$). Se observaron diferencias importantes entre las distintas vías de administración del cloruro de paladio (II), poniéndose de manifiesto en ratas Charles-River CD1 valores para la DL_{50} de 5 mg/kg de peso corporal en la vía intravenosa, 6 mg/kg de peso corporal en la intratraqueal, 70 mg/kg de peso corporal en la intraperitoneal y 200 mg/kg de peso corporal en la oral. En ratas Sprague-Dawley se obtuvieron valores más altos para la DL_{50} por vía oral.

Los signos de la toxicidad aguda observados con varias sales de paladio en ratas o conejos fueron la muerte, una reducción de la ingesta de alimentos y de agua, emaciación, casos de ataxia y marcha de puntillas, convulsiones clónicas y tónicas, efectos cardiovasculares, peritonitis o cambios bioquímicos (por ejemplo, cambios en la actividad de las enzimas hepáticas, proteinuria o cetonuria). Se detectaron cambios funcionales o histológicos tanto con los compuestos de paladio como con el polvo de paladio elemental. También se produjeron hemorragias en los pulmones y en el intestino delgado.

Los efectos registrados en roedores y conejos tras una exposición breve a varios compuestos de paladio corresponden principalmente a cambios en parámetros bioquímicos (por ejemplo, disminución de la actividad de las enzimas microsomales hepáticas o de la producción de proteínas microsomales). Los signos clínicos fueron inactividad, pérdida de peso, hematomas o exudaciones. También se observaron cambios en el peso absoluto y relativo de los órganos y anemia. Un compuesto (el complejo tetracloropaladato (II) de sodio - albúmina de huevo) provocó la muerte de ratones. Las concentraciones efectivas eran del orden de varios mg por kg de peso corporal. Se detectaron efectos histopatológicos en el hígado, el riñón, el bazo o la mucosa gástrica de ratas 28 días después de la administración diaria por vía oral de 15 ó 150 mg de bicarbonato de paladio tetraamina ($[Pd(NH_3)_4](HCO_3)_2$)/kg de peso corporal. Además, se produjo un aumento del peso absoluto del cerebro y los ovarios con dosis de 1,5 y 15 mg/kg de peso corporal.

No está clara la contribución del paladio a los efectos observados tras la administración única o breve de una aleación dental con paladio.

También se dispone de pocos datos sobre los efectos de la exposición prolongada a diversas especies (formas) de paladio.

En ratones tratados con cloruro de paladio (II) (5 mg de paladio/l) en el agua de bebida desde el destete hasta la muerte natural se puso de manifiesto una disminución del peso corporal, un ciclo de vida más largo (en los machos, pero no en las hembras), un aumento en la amiloidosis de varios órganos internos y alrededor del doble de tumores malignos (véase *infra*).

La exposición de ratas por inhalación a cloropaladosamina (($NH_3)_2PdCl_2$) durante alrededor de seis meses provocó cambios ligeros reversibles (a 5,4 mg/m^3) o significativos y permanentes (a 18 mg/m^3) en varios parámetros del suero sanguíneo y la orina, que indicaban daños principalmente del hígado y el riñón (además de una reducción del aumento del peso corporal, cambios en el peso de los órganos y glomerulonefritis). También se detectaron efectos adversos con exposiciones entéricas, siendo la concentración sin efectos adversos observados de 0,08 mg/kg de peso corporal.

Seis meses después de una aplicación intratraqueal única de polvo de paladio (143 mg/kg de peso corporal) se observaron varios signos histopatológicos de inflamación en los pulmones de ratas. La administración diaria por vía oral de polvo de paladio (50 mg/kg de peso corporal) durante seis meses provocó cambios en varios parámetros del suero sanguíneo y la orina de ratas.

Las pruebas cutáneas con conejos de una serie de compuestos de paladio pusieron de manifiesto reacciones cutáneas de distinta gravedad, lo que ha permitido establecer la siguiente clasificación: $(NH_4)_2PdCl_6 > (NH_4)_2PdCl_4 > (C_3H_5PdCl)_2 > K_2PdCl_6 > K_2PdCl_4 > PdCl_2 > (NH_3)_2PdCl_2 > PdO$. Los tres primeros compuestos provocaron eritema, edema o escara en la piel intacta y escarificada, los tres siguientes eritema en la piel escarificada y los dos últimos no fueron irritantes. El clorhidrato de paladio (no se ha facilitado la fórmula) también produjo dermatitis en la piel de conejos.

Se observó irritación ocular en conejos tras la aplicación de cloruro de paladio (II) y bicarbonato de paladio tetraamina (pero no con el óxido de paladio (II)) en la superficie de los ojos. La exposición por inhalación a la cloropaladosamina (≥ 50 mg/m^3) afectó a las membranas mucosas de los ojos de ratas (conjuntivitis, querato-conjuntivitis).

Se ha observado que algunos compuestos de paladio son potentes sensibilizadores cutáneos (cloruro de paladio (II), bicarbonato de paladio tetraamina, clorhidrato de paladio [no se ha especificado la fórmula], complejos de paladio-albúmina). En una prueba de maximización con cobayas se puso de manifiesto que el cloruro de paladio (II) era un sensibilizador más potente que el sulfato de níquel (NiSO$_4$). Los cobayas sensibilizados con cromato o sales de cobalto o níquel no reaccionaron tras la aplicación de cloruro de paladio (II). Sin embargo, si se sensibilizaban con cloruro de paladio (II), reaccionaban frente al sulfato de níquel. Se han obtenido resultados algo divergentes en pruebas en las que se estudiaba la reactividad cruzada entre el paladio y el níquel mediante aplicaciones abiertas repetidas en la piel de cobayas. En estos experimentos se sensibilizaron los animales con cloruro de paladio (II) ($n = 27$) o sulfato de níquel ($n = 30$) con arreglo al método de prueba de maximización de cobayas y luego se trataron una vez al día durante 10 días de acuerdo con la prueba de aplicaciones abiertas repetidas mediante la administración de alergeno sensibilizante (cloruro de paladio (II) o sulfato de níquel), así como del compuesto que puede producir una posible reacción cruzada (sulfato de níquel o cloruro de paladio (II)) y el vehículo tópico en cobayas. En este estudio siguió sin quedar claro si la reactividad al cloruro de paladio (II) en animales sensibilizados con sulfato de níquel se debía a una reactividad cruzada o a la inducción de sensibilidad por los tratamientos repetidos. Por otra parte, la reactividad frente al sulfato de níquel en los animales sensibilizados con cloruro de paladio (II) podría atribuirse a una reacti-vidad cruzada. Se ha observado en gatos sensibilización respiratoria (broncoespasmos) tras la administración intravenosa de varios com-puestos complejos de paladio. Iba acompañada de un aumento de la concentración de histamina en el suero. Se han obtenido respuestas inmunitarias significativas con cloruro de paladio (II) y/o cloro-paladatos utilizando la valoración de los nódulos linfáticos poplíteos y auriculares en ratones BALB/c. Los datos preliminares obtenidos de un modelo de animales parecen indicar que los compuestos de paladio (II) podrían intervenir en la inducción de una enfermedad autoinmunitaria.

No hay datos suficientes sobre los efectos del paladio y sus compuestos en la reproducción y el desarrollo. En un estudio de detección se notificó un peso reducido de los testículos en los ratones que habían recibido 30 dosis subcutáneas diarias de cloruro de paladio (II), con una dosis total de 3,5 mg/kg de peso corporal.

Puede haber una interacción de compuestos de paladio con ADN aislado *in vitro*. Sin embargo, con una sola excepción, las pruebas de mutagenicidad de varios compuestos de paladio con células de bacterias o de mamíferos *in vitro* (prueba Ames: *Salmonella typhimurium*; prueba cromática SOS: *Escherichia coli*; prueba del micronúcleo: linfocitos humanos) dieron resultados negativos. Asimismo, en una prueba de genotoxicidad *in vivo* (prueba del micronúcleo en el ratón) con bicarbonato de paladio tetraamina se obtuvieron resultados negativos.

Se han notificado tumores asociados con la exposición al paladio en dos estudios. Los ratones tratados con cloruro de paladio (II) (5 mg de Pd^{2+}/l) en el agua de bebida desde el destete hasta la muerte natural contrajeron tumores malignos, fundamentalmente de los tipos linfoma-leucemia y adenocarcinoma del pulmón, con una tasa estadísticamente significativa, pero coincidiendo con una mayor longevidad en los machos, que puede explicar por lo menos en parte el aumento del número de tumores. Se observaron tumores en el lugar de implantación en 7 de 14 ratas (no estaba claro si los tumores se debían al estímulo físico crónico o a los compuestos químicos) 504 días después de la implantación subcutánea de una aleación de plata-paladio-oro. No había ningún estudio de la carcinogenicidad para la exposición por inhalación.

Los iones de paladio pueden inhibir la mayor parte de las funciones celulares importantes, como se ha observado *in vivo* e *in vitro*. El punto más sensible parece ser la biosíntesis de ADN/ARN. El valor de la CE_{50} del cloruro de paladio (II) para la inhibición de la síntesis de ADN *in vitro* con fibroblastos de ratón fue de 300 µmol/l (32 mg Pd^{2+}/l). En ratas tratadas con una dosis intraperitoneal única de 14 µmol de nitrato de paladio (II) ($Pd(NO_3)_2$)/kg de peso corporal (1,5 mg Pd^{2+}/kg de peso corporal) se produjo la inhibición de la síntesis de ADN *in vivo* (en el bazo, el hígado, el riñón y los testículos).

Cuando se evaluó microscópicamente el paladio aplicado en su forma metálica se observó una citotoxicidad *in vitro* nula o pequeña.

Se ha comprobado que las sales de paladio simples y complejas inhiben una serie de enzimas aisladas con funciones metabólicas básicas. La mayor inhibición (valor de la K_i para el cloruro de paladio (II) = 0,16 µmol/l) se detectó para la creatinina kinasa, enzima importante del metabolismo energético.

Numerosos complejos orgánicos de paladio tiene un potencial antineoplásico semejante al del *cis*-dicloro-2,6-diaminopiridina-platino (II) (*cis*-platino, medicamento anticanceroso).

El mecanismo de acción de los iones de paladio y del paladio elemental no está totalmente claro. La formación de complejos de los iones de paladio con componentes celulares probablemente desempeña inicialmente una función básica. Podrían intervenir asimismo procesos de oxidación, debido a los diferentes estados de oxidación del paladio.

7. Efectos en el ser humano

No hay información sobre los efectos en la población general de las emisiones de paladio procedentes de los catalizadores de los automóviles. Se han notificado efectos debido a exposiciones iatrogénicas y de otro tipo.

La mayoría de los casos notificados se refieren a la sensibilidad al paladio asociada con la exposición a arreglos dentales con aleaciones que contienen paladio, cuyos síntomas son dermatitis por contacto, estomatitis o mucositis y liquen de Wilson oral. Los pacientes con pruebas del parche positivas al cloruro de paladio (II) no reaccionaban necesariamente al paladio metálico. Sólo algunas personas que dieron resultado positivo a la prueba del parche con cloruro de paladio (II) mostraron síntomas clínicos en la mucosa oral como consecuencia de la exposición a aleaciones con paladio. En un estudio se observaron cambios ligeros, pero no significativos, en las inmunoglobulinas del suero tras un arreglo dental con una aleación de plata-paladio.

Los efectos secundarios de las preparaciones de paladio observados en otros usos médicos o experimentales incluyen fiebre, hemólisis, discoloración o necrosis en los lugares de inyección tras la administración subcutánea y eritema y edema después de la aplicación tópica.

En un pequeño número de informes se describieron casos de trastornos cutáneos en pacientes que habían estado expuestos a joyas que contenían paladio o a fuentes no especificadas.

En pruebas del parche seriadas con cloruro de paladio (II) se puso de manifiesto una alta frecuencia de sensibilidad al paladio en grupos especiales objeto de estudio. En varios estudios recientes y de gran tamaño de diferentes países se encontraron frecuencias de sensibilidad al paladio del 7 al 8% en pacientes de clínicas dermatológicas, así como en escuelas, con predominio en las mujeres y en las personas más jóvenes. En comparación con otros alergenos (se estudiaron unos 25), el paladio está entre los siete sensibilizadores que con más frecuencia provocan reacción (clasificado en segundo lugar tras el níquel dentro de los metales). Se observaron reacciones aisladas al paladio (mono-alergia) con una frecuencia baja. Fundamentalmente se han detectado reacciones combinadas con otros metales (multisensibilidad), en particular con el níquel.

Hasta ahora, las fuentes de sensibilización al paladio identificadas con mayor frecuencia para la población general son los arreglos dentales y la joyería.

Hay pocos datos sobre los efectos adversos en la salud debidos a la exposición ocupacional al paladio. Unos pocos trabajadores de metales del grupo del platino (2/307; 3/22) dieron reacción positiva a una sal compleja de haluro de paladio en pruebas de sensibilización (prueba de puntura cutánea; prueba del radioalergoabsorbente; prueba de anafilaxis cutánea pasiva en monos). Algunos trabajadores (4/130) de una fábrica de catalizadores de automóvil dieron reacciones positivas en las pruebas de puntura con cloruro de paladio (II). En un estudio analítico (sin detalles) se informaba de la aparición frecuente de enfermedades alérgicas de las vías respiratorias, dermatosis y afecciones de los ojos entre trabajadores rusos de la producción de metales del grupo de platino. Se han documentado tres casos aislados de dermatitis alérgica por contacto en dos químicos y un trabajador del metal. En la industria electrónica se ha observado un caso aislado de asma ocupacional inducida por sales de paladio.

Las subpoblaciones con riesgo especial de alergia al paladio son las personas con alergia conocida al níquel.

8. Efectos en otros organismos en el laboratorio y en el medio ambiente

Se ha observado que varios compuestos de paladio tienen propiedades antivíricas, antibacterianas y/o fungicidas. Raramente se han realizado pruebas normalizadas de toxicidad microbiana en condiciones ecológicamente adecuadas. Se ha obtenido una CE_{50} en 3 horas de 35 mg/l (12,25 mg de paladio/l) para el efecto inhibitorio del bicarbonato de paladio tetraamina en la respiración de lodos cloacales activados.

Se ha observado que los compuestos de paladio sometidos a prueba para analizar los efectos en los organismos acuáticos tienen una toxicidad significativa. Dos complejos de paladio (tetracloropaladato (II) de potasio y cloropaladosamina) presentes en soluciones nutritivas provocaron necrosis en el jacinto de agua (*Eichhornia crassipes*) con 2,5-10 mg de paladio/l. La toxicidad aguda (CL_{50} a las 96 horas) del cloruro de paladio (II) para el anélido tubícola de agua dulce *Tubifex tubifex* fue de 0,09 mg de paladio/l. Se ha notificado para el pez de agua dulce medaka (*Oryzias latipes*) una concentración letal mínima en 24 horas de 7 mg de cloruro de paladio (II)/l (4,2 mg de paladio/l). En todos los casos, los compuestos de paladio tenían una toxicidad similar a la de los compuestos de platino.

Sólo se han efectuado las pruebas de toxicidad en organismos acuáticos de acuerdo con las directrices de la Organización de Cooperación y Desarrollo Económicos para el bicarbonato de paladio tetraamina. Se obtuvo un valor de la CE_{50} a las 72 horas de 0,066 mg/l (equivalentes a 0,02 mg de paladio/l) (prueba de la inhibición de la multiplicación celular con *Scenedesmus subspicatus*), una CE_{50} a las

48 horas de 0,22 mg/l (0,08 mg de paladio/l) (inmovilización de *Daphnia magna*) y una CL_{50} a las 96 horas de 0,53 mg/l (0,19 mg de paladio/l) (toxicidad aguda para la trucha arco iris *Oncorhynchus mykiss*). Se obtuvieron unas concentraciones sin efectos observados (NOEC) de 0,04 mg/l (0,014 mg de paladio/l) (algas), 0,10 mg/l (0,05 mg de paladio/l) (*Daphnia magna*) y 0,32 mg/l (0,11 mg de paladio/l) (peces). Todos estos valores se han basado en concentraciones nominales. Sin embargo, con frecuencia se ha encontrado que las concentraciones medidas correspondientes eran mucho más bajas y variables; no están claras las razones de esto. Se han calculado valores basados en las concentraciones medias medidas ponderadas por el tiempo para la prueba de inmovilización con *Daphnia magna*, con una CE_{50} a las 48 horas de 0,13 mg/l (0,05 mg de paladio/l) y una NOEC de 0,06 mg/l (0,02 mg de paladio/l). Se han observado asimismo efectos fitotóxicos en las plantas terrestres tras la adición de cloruro de paladio (II) a la solución nutritiva. Son inhibición de la transpiración a 3 mg/l (1,8 mg de paladio/l), cambios histológicos a 10 mg/l (6 mg de paladio/l) o la muerte a 100 mg/l (60 mg de paladio/l) en la poa (*Poa pratensis*). En algunos cultivos de plantas se produjo un retraso del crecimiento y atrofia de las raíces dependientes de la dosis, siendo la avena la más sensible, que se vio afectada por unos 0,22 mg de cloruro de paladio (II)/l (0,132 mg de paladio/l).

No se ha encontrado información bibliográfica sobre los efectos del paladio en los invertebrados o los vertebrados terrestres.

No se dispone de observaciones en el medio ambiente.

THE ENVIRONMENTAL HEALTH CRITERIA SERIES(continued)

Flame retardants: tris(chloropropyl) phosphate and tris(2-chloroethyl) phosphate (No. 209, 1998)
Flame retardants: tris(2-butoxyethyl) phosphate, tris(2-ethylhexyl) phosphate and tetrakis(hydroxymethyl) phosphonium salts (No. 218, 2000)
Fluorine and fluorides (No. 36, 1984)
Food additives and contaminants in food, principles for the safety assessment of (No. 70, 1987)
Formaldehyde (No. 89, 1989)
Fumonisin B_1 (No. 219, 2000)
Genetic effects in human populations, guidelines for the study of (No. 46, 1985)
Glyphosate (No. 159, 1994)
Guidance values for human exposure limits (No. 170, 1994)
Heptachlor (No. 38, 1984)
Hexachlorobenzene (No. 195, 1997)
Hexachlorobutadiene (No. 156, 1994)
Alpha- and beta-hexachlorocyclohexanes (No. 123, 1992)
Hexachlorocyclopentadiene (No. 120, 1991)
n-Hexane (No. 122, 1991)
Human exposure assessment (No. 214, 2000)
Hydrazine (No. 68, 1987)
Hydrogen sulfide (No. 19, 1981)
Hydroquinone (No. 157, 1994)
Immunotoxicity associated with exposure to chemicals, principles and methods for assessment (No. 180, 1996)
Infancy and early childhood, principles for evaluating health risks from chemicals during (No. 59, 1986)
Isobenzan (No. 129, 1991)
Isophorone (No. 174, 1995)
Kelevan (No. 66, 1986)
Lasers and optical radiation (No. 23, 1982)
Lead (No. 3, 1977)[a]
Lead, inorganic (No. 165, 1995)
Lead – environmental aspects (No. 85, 1989)
Lindane (No. 124, 1991)
Linear alkylbenzene sulfonates and related compounds (No. 169, 1996)
Magnetic fields (No. 69, 1987)
Man-made mineral fibres (No. 77, 1988)
Manganese (No. 17, 1981)
Mercury (No. 1, 1976)[a]
Mercury – environmental aspects (No. 86, 1989)
Mercury, inorganic (No. 118, 1991)
Methanol (No. 196, 1997)
Methomyl (No. 178, 1996)
2-Methoxyethanol, 2-ethoxyethanol, and their acetates (No. 115, 1990)
Methyl bromide (No. 166, 1995)
Methylene chloride
(No. 32, 1984, 1st edition)
(No. 164, 1996, 2nd edition)
Methyl ethyl ketone (No. 143, 1992)
Methyl isobutyl ketone (No. 117, 1990)
Methylmercury (No. 101, 1990)
Methyl parathion (No. 145, 1992)

Methyl tertiary-butyl ether (No. 206, 1998)
Mirex (No. 44, 1984)
Morpholine (No. 179, 1996)
Mutagenic and carcinogenic chemicals, guide to short-term tests for detecting (No. 51, 1985)
Mycotoxins (No. 11, 1979)
Mycotoxins, selected: ochratoxins, trichothecenes, ergot (No. 105, 1990)
Nephrotoxicity associated with exposure to chemicals, principles and methods for the assessment of (No. 119, 1991)
Neurotoxicity associated with exposure to chemicals, principles and methods for the assessment of (No. 60, 1986)
Neurotoxicity risk assessment for human health, principles and approaches (No. 223, 2001)
Nickel (No. 108, 1991)
Nitrates, nitrites, and N-nitroso compounds (No. 5, 1978)[a]
Nitrogen oxides
(No. 4, 1977, 1st edition)[a]
(No. 188, 1997, 2nd edition)
2-Nitropropane (No. 138, 1992)
Noise (No. 12, 1980)[a]
Organophosphorus insecticides: a general introduction (No. 63, 1986)
Palladium (No. 226, 2001)
Paraquat and diquat (No. 39, 1984)
Pentachlorophenol (No. 71, 1987)
Permethrin (No. 94, 1990)
Pesticide residues in food, principles for the toxicological assessment of (No. 104, 1990)
Petroleum products, selected (No. 20, 1982)
Phenol (No. 161, 1994)
d-Phenothrin (No. 96, 1990)
Phosgene (No. 193, 1997)
Phosphine and selected metal phosphides (No. 73, 1988)
Photochemical oxidants (No. 7, 1978)
Platinum (No. 125, 1991)
Polybrominated biphenyls (No. 152, 1994)
Polybrominated dibenzo-p-dioxins and dibenzofurans (No. 205, 1998)
Polychlorinated biphenyls and terphenyls
(No. 2, 1976, 1st edition)[a]
(No. 140, 1992, 2nd edition)
Polychlorinated dibenzo-p-dioxins and dibenzofurans (No. 88, 1989)
Polycyclic aromatic hydrocarbons, selected non-heterocyclic (No. 202, 1998)
Progeny, principles for evaluating health risks associated with exposure to chemicals during pregnancy (No. 30, 1984)
1-Propanol (No. 102, 1990)
2-Propanol (No. 103, 1990)
Propachlor (No. 147, 1993)
Propylene oxide (No. 56, 1985)
Pyrrolizidine alkaloids (No. 80, 1988)
Quintozene (No. 41, 1984)
Quality management for chemical safety testing (No. 141, 1992)

---

[a] Out of print